Icing on the Damper

To Allan
Christmas '96
from John & Di
with love.

Marie Mahood

Icing on the Damper

Life Story of a Family
in the Outback

Central Queensland
UNIVERSITY
PRESS

© Marie Mahood 1995

First published 1995 by
CQU Press
PO Box 1615
Rockhampton Q 4700
Tel: 079-228 144
Fax: 079-228 151

Reprinted in 1996

National Library of Australia
Cataloguing in Publication entry

Mahood, Marie

Icing on the Damper: life story of a family
      in the outback

ISBN 1 875998 03 9

1. Mahood family. 2. Ranch life - Northern
Territory. 3. Ranch life - Queensland.
4. Frontier and pioneer life - Northern
Territory. I. Title.

636.201099429

Typeset by Anne Camilleri
in Barret 12 pt and Tango 36 pt.

Printed and bound by
Watson Ferguson,
Brisbane, Qld.

Cover Painting *West of Alice* by Joe Mahood, 1989.

> *. . . I bought the only white dress in the hawker's van, altered by the town dressmaker, who also lent us her wedding ring for the ceremony. Someone located some roses and made a bouquet; someone else iced the Big Sister fruit cakes to perfection. The government grader driver with the only film and flashlight in town took our wedding photos for us.*

Joe Mahood and Marie Healy, married at Katherine, N.T. on 10 June 1952.

Years later in Queensland, the Mahoods remained a close-knit family. Here is how they handled cleanskins.

... *Branding was a team effort. Harry lassoed the cleanskins and big weaners and tied the end of the rope to the Toyota bullbar. I backed the Toyota away to drag the beast up to the bronco panel. With Harry on the front leg and Kim and Bob each grabbing a back leg Joe applied the brands and castrated the males. Tracey passed the brands and kept the fire hot and Jimmy did the ear-marking.*

The Mahood Family at Mongrel Downs Cartoon by Joe Mahood, 1962.

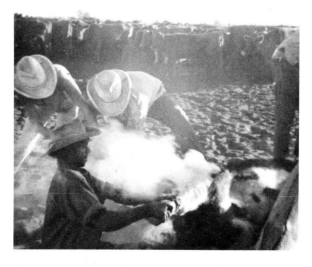

Open-bronco branding at Mongrel Downs, 1970.

For my grandchildren

ARTS QUEENSLAND

Sponsored by the Queensland Office of Arts and Cultural Development

This project has been assisted by the Commonwealth
Government through the Australia Council, its arts
funding and advisory body.

CONTENTS

Chapter 1

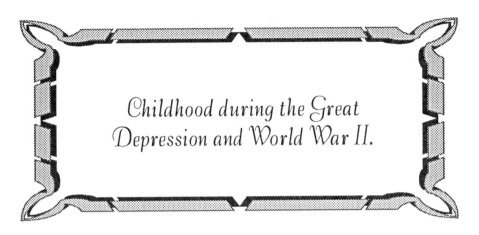

Childhood during the Great
Depression and World War II.

From horizon to horizon there was nothing but spinifex and the occasional squat red-brown anthill or low acacia bush, bisected by a two-wheel sandy red track. It was slow going, four-wheel drive, in the battered old Landrover. Joe was mildly surprised when he saw the two Aborigines, an older man and a youth, carrying spears and a billy-can, strolling along the track ahead of him. He pulled up and offered a lift though the track led nowhere but to two abandoned gold-mining sites. The old man declined, they were on blackfella business, but he asked to top up his billy-can from Joe's water canteen. "You gottim smoke, boss?" Joe rolled them each a cigarette and asked conversationally, "How far to Tanami?" He'd never travelled this track before and the old chap was obviously familiar with the country.

The alert brown eyes in the leathery face twinkled. "You gotta footwalk — 'im lo-ong way. You gottim motor-car — 'im littlebit long way. You gottim aeroplane — by Krise, boss, 'im 'ere now!"

Joe deduced he had about forty miles to go, thanked them and drove on.

That's the story of our life together, aiming for the goal of a cattle station of our own, a mighty lot of footwalking, now and then a motor-car ride, and a couple of strokes of luck, being in the right place at the right time, which I suppose you could equate to an aeroplane ride.

Joe was born Alexander James Mahood on the 20th of March, 1928, at Flemington, near Sydney, second child in a family of five. His ex-Navy dad was soon to become one of the victims of the Great Depression, two days' work a week on the roads and supply your own pick and shovel. I was born on the 9th of May, 1926, on the other side of the continent at Victoria Park, Perth, elder of two. My dad, just starting his one-man carrying business, had his truck re-possessed when he missed a week's hire-purchase payment because he was hospitalised with pneumonia. When he recovered, my parents sold their few bits and pieces and Mum, pregnant with my younger sister, took me back to her parents' home in the country to start a boarding house while Dad bought a tent and the basic gear to go prospecting on the goldfields out past Kalgoorlie. Sadly, Mum and Dad never did get together again, though they always remained friends. The Depression broke up a lot of families; perhaps the war which was soon to follow broke up even more.

Joe was a serious little boy — a bit of a loner. His family called him "The Dreamer". It was obvious very early that he was a gifted artist and he always topped his class at school, despite the fact that, from the age of ten, he held down a job driving a milk cart and delivering milk from 1 until 8 in the morning, then a quick breakfast and a long walk to school.

On Saturdays he put in the full day at the dairy, grooming the horses, washing down the carts, cleaning cans and hosing out. For this he was paid a shilling a night and free milk for the family.

On Sundays he explored the surrounding bushland, sketched, or swam and canoed in the Georges River with his brothers and sister. He dreamed about riding spirited horses, going droving in the outback, and he and a mate, Johnny Ashcroft, acquired guitars in their early teens and, self-taught, played and sang the bush ballads of the twenties and thirties.

Aged 14, dux of the top-stream of the Grade 10 "Junior" year at Granville Technical School, Joe's student aptitude test suggested training as a fitter and turner; so did the tests of almost every other boy who had achieved a passing grade. Australia was at war and Fitting and Turning was a premium trade in the munitions factories and engineering works. Joe's almost 100% marks in English, History, Geography and Art as well as the Maths, Sciences and Technical subjects had no bearing on the advice, which obviously should have been to continue on to tertiary education.

"The Dreamer" could not envisage his wide and limitless horizons trimmed down to the four walls of a noisy workshop. Aged just fifteen, well-built but small for his age, he put together a couple of blankets and a change of clothes, a writing pad and his beloved guitar and headed north for Queensland and the cattle country.

On the way he worked at various harvest-time farm jobs, refused offers of permanent work, occasionally jumped trains, and as the months went by, his objective grew closer. His will had only once faltered, early in his odyssey, when he had earned at a job on an arrowroot farm the first money he could really call his own and spend as he wished.

He decided his unkempt appearance merited a visit to the barber. He had never been to a barber's shop in his life because his mother had always previously cut his hair. In those Depression days the luxury of professional hairdressers was far beyond the means of any working-class family.

His two pounds cash in his pocket, Joe entered the barber shop in a Northern NSW coastal town. The barber called him Sir. "Not just a trim, Sir. You really need a shampoo-treatment, Sir. And this, Sir, and that, Sir." Joe was rich; he ordered the lot. When the barber flicked the towel away from his shoulders and said, "That'll be twenty-five shillings, Sir!" Joe came back to earth with a thud.

In his wildest estimation he had thought it might cost him five shillings — now he had only fifteen shillings left, and train fare north and food to buy. He had planned on a meal and a room for the night in a boarding house, but now he felt so sick with disgust at himself that he wasn't even hungry — he couldn't afford to be. He carried his swag to the wooded hillside above the town and lay on his blankets and watched the dusk fall, saw the smoke rise from chimneys as fires were lit to prepare the evening meal, heard the bark of a dog which sounded just like the bark of the family dog at home. One by one the lights came on in cottage windows and he began to picture the families gathering around their meal-tables and, as a natural extension, his own family in their warm kitchen, laughing and eating and saving the chop bones for the old dog crouched expectantly just outside the kitchen door.

With full dark it grew bitterly cold and he realised that he was hungry. He turned his back on the lights that projected the comfort and warmth of home and played a few nostalgic tunes on his guitar to take his mind off his growling stomach and the almost unbearable ache for his own home and family.

Joe grinned later when he told me how the barber must have assessed the raw youth and made his killing but, as far as I know, Joe never went into a hairdresser's shop again in his life and he always washed his hair with soap instead of shampoo. He let his hair grow long many years before short back and sides went out of fashion, and when it drew too much adverse comment anyone who offered could take the scissors to it, but mostly he cut it himself with a razor blade. For our sons' weddings he let me trim the tufts he'd missed at the back where he couldn't see.

By the time he was sixteen he had achieved the first step in his ambition and was working as a stockman (in the cattle country the term is "ringer") on a property in the Mt.Isa district.

In the same year, I was completing my final year at a Perth boarding school to which I had won a scholarship. My aptitude test rated me as a potential commercial artist. Well, that was all right, but I had long ago decided that I was going to be a writer.

At eight I had written my 200-page-length first novel, entitled "Bobby" or "Jungle Terrors", later to be used to light the copper fire by my prosaic mother who didn't hold with such romantic day-dreams, but at the same time worked her guts out running the boarding-house and dress-making so that my younger sister and I could get a good education in preparation for a professional job with regular wages, and therefore an escape from the prospect of a lifetime beholden to some man for succour and support.

Mum was an early feminist, though not quite as strident about it as her own mother, my fierce grandmother, who disapproved of all men in general and Irish men in particular. Grandma could quote an adage or a proverb to suit any occasion. My own father she had early dismissed as a jack of all trades and master of none, a rolling stone who gathered no moss, and one who would never wake up to the fact that a bird in the hand was worth two in the bush. As she went about her work she used to hum a little song, the refrain of which went:

"Girls keep away from the boys,
Give them whips of room,
For you'll find when you're wed
They'll hit you on the head
With the bald-headed end of a broom."

She disapproved heartily of my compulsive reading habit — "just like her good-for-nothing father!" — and registered the opinion that I'd come to a bad end if I didn't pull my socks up.

When I took a job as offsider to the editor of a weekly newspaper she said, "Well, she's made her bed; let her lie on it!". She added that it was now inevitable that, having chosen to stray from the straight and narrow path of respectable employment, I'd end up spending most of my life tagging after some *man*. I'm jolly glad I did; we had a lot more ups than downs.

Joe and I met in 1950 on Victoria River Downs, then a property of 13,000 square miles, the world's largest cattle station. I'd spent four years at University in Perth on a Hackett Bursary, doing an Arts degree and an Honours course in English, and then worked for a year as a reporter on *The West Australian*.

On the spur of the moment I'd decided to freelance, and Uncle Jack's one-man police station at Halls Creek in the Kimberleys was a good jumping-off-place for the frontier North about which I intended to write. I thought I knew the lot, the sophisticated keen young reporter, when I stepped off the little plane into the searing Christmas Eve heat of Halls Creek, population about twenty during the year, but swollen to three times that number for the festive celebrations.

Chapter 2

We Meet on Victoria River Downs and
Get Married at a Picnic Race Meeting.

Inside a month, I realised that my education was only just beginning. It was a tough country — raw, basic, larger than life, and so were all the people in it. It was all I expected and more.

The people I met were "do-ers," not just the "talkers" of my University days, and, to my avid ears, they were better talkers too, with the colourful metaphor and tongue-in-cheek exaggeration of the outback.

Here, the art of conversation was alive and lusty with horizons not bounded by the restrictions of white middle-class suburbia, all I'd previously known when I'd mixed only with people of similar background to myself. Two months later I had taken a job in the store at Victoria River Downs where three other girls were employed as book-keeper, clerk, and teacher for the eight or so children of the manager and white staff.

If the small, family-owned stations of the north were examples of free-and-easy social intercourse where a person was judged only on character and ability, Victoria River Downs was the exact opposite. Like all the English company-owned great stations of the time, Victoria River Downs operated under a feudal system, the like of which hadn't been seen in England for centuries.

Right at the top was Lord Luke back in England, owner of the Bovril Empire, who was on a very good thing because until 1950 the company had paid no income tax in Australia. On the 13,000 square-mile Australian demesne, the manager ruled supreme from The Big House, with his private chef and retinue of black servants.

Important social visitors were entertained at The Big House but run-of-the-mill employees on station business never got further than the garden path at the bottom of the steps. The manager stood on the top step and talked down.

Next in the hierarchy came the overseer and the girls. We ate in the dining room, together with less important visitors on mere business and any of the visiting head-stockmen from the five outstations. Our cook had his offsiders, including a little black boy who pulled a punkah while we dined, but the cook didn't eat with us; his place was in the kitchen with the other single station hands.

Every Saturday night he got time off when the overseer and the girls attended a command meal at The Big House. Married employees lived a more normal life in their cottages. The white ringers and station hands were "kitchen" class; we girls were officially instructed not to mix with them socially.

As each stock-camp had its own cook, the homestead camp, Centre Camp, on those rare occasions when it was in at the station, was quartered in a big tin shed and fed by its own cook on the usual beef, damper, spuds and onions. Joe was a ringer in Centre Camp.

At this time the employees were paid a wage that didn't include sick pay, workers' compensation or holiday pay. If they were injured on the job, bad luck! The Aborigines were the serfs at the bottom of the heap. They worked for their tucker and clothes; nobody gave a thought to any schooling for their children.

That was the first year payment for Aboriginal workers was introduced, a pound a week for stockmen and ten shillings a week for house-girls but, as they now had to buy their own clothes, blankets and tobacco, nothing had really changed much for them.

I didn't give it too much thought then, apart from a steadfast refusal to keep my appointed place and a vague pity for the seldom-seen wife of the manager in her lonely isolation at the top. With the continual stream of customers in the store, station personnel, drovers and travellers, I soon learned a great deal more about the colourful people of the outback. In quiet moments Peter, the elderly Aboriginal store employee, told me many fascinating stories about his people and their culture.

It was the custom, on Sundays, for the overseer to escort us girls on horseback to various picnic spots or to visit the stock-camp at times when the men were horse-breaking or mustering within a day's ride. I first took extra notice of Joe on one such day when the fine-looking big black colt he was riding suddenly began to buck furiously and he gave us a superb exhibition of riding until the overseer sourly motioned us away. Joe laughingly told me afterwards that he'd tickled the horse up with his offside spur to make it buck so he could show off to the girls, a fact the overseer was probably well aware of. It had been a stirring demonstration.

Then came the Saturday royal invitation to dine that I declined, much to the horror of the other girls. "You can't!" they gasped. "I can," I said. "Three or four hours of boredom being patronised by a man I actually despise isn't worth a flasher meal than usual. What's more, I'm not going next week either!" I think they told him I was sick. I settled down happily with a cup of tea, a packet of biscuits and a sketchbook.

I didn't know that the Centre Camp stock-camp had arrived with a mob of cattle late that afternoon but when Joe, seeing the light in our cottage, rode over, I was pleased with the company. He idly picked up my pencil and began to sketch when I told him I could never get the horses' legs quite right. It was magic. In a few swift lines horses and then men came to life on the page.

He gave a little sideways grin and there we four girls were too. I was delighted. I offered Joe a drink from the hidden bottle of creme-de-menthe a drover had surreptitiously brought me from Wyndham. The upshot was that we drank the whole bottle and giggled and kissed each other lengthily over the back fence, where his horse patiently waited. Then he unhitched the reins, mounted on the nearside and fell off the offside. On the second attempt he stayed in the saddle and the horse ambled off in the moonlight. I told myself as I walked back down the pathway, hiccupping slightly, that an evening drinking creme-de-menthe with Joe knocked spots off an evening sipping sherry with the manager.

Two evenings later the overseer escorted us four girls on horseback to have a look at the mob of cattle awaiting delivery to a drover and to eat a picnic tea in Centre Camp. After we ate, Joe was persuaded to bring out his guitar as we sat around the campfire. A warm, starry night, the dancing firelight, the occasional soft lowing of cattle in the background and the moving silhouettes of the two horsemen on night-watch. Add to that a rich tenor voice singing on demand all the evocative ballads of the Australian outback.

I was entranced. I'd been told that Joe was a guitarist and singer of some repute but this was a revelation. This lithe young ringer in the dark blue open-necked shirt and the tight blue stockman-cut pants, with a lock of hair falling over his forehead as he played, suddenly became the centre of all my attention. He looked up and caught my glance. Something like a bolt of lightning seemed to pass between us. He plucked the guitar strings and held my eyes as he sang softly the song that began,

I'd climb the highest mountain
If when I climbed that mountain
I'd find you.
I'd swim the deepest river
If when I swam that river
I'd find you.

Then he stood up, walked around the campfire and helped me to my feet, saying, "Wait until I put my guitar away and get one of the other fellows to stand in for my watch and I'll ride back to the station with you".

That ride in the moonlight, holding hands as we rode together, sealed both our futures.

"I never gave a thought to getting married until about an hour ago," Joe said dreamily.

"Me either, not seriously," I said.

"I suppose we'd better wait until I've saved enough for a good start."

"I'll save too!"

We were delirious with the magic of it. True, I'd been proposed to before — by a city academic, a Swedish migrant, a Communist organiser, and the last one a drover who'd told me I'd be worth six dogs in a man's swag. But none of them could sketch like Rembrandt, sing like Caruso, ride like Darby Munro and cause little lightnings to leap from his fingertips when he merely touched my arm. In short, I loved him.

"I love you, Honey," he said, "so much — more than too much! I'd fight my weight in wildcats for you! And lick 'em easily, too!"

"Well, I would too for you," I began, then had a second thought, "maybe one — if it was a little one — and I had a stick!"

We laughed together and kissed and parted.

The next day Centre Camp handed over the drover's mob and went bush again. Two days later the manager told me to grease my swag straps and take the next plane out. In other words, I was sacked. It wasn't unexpected. At the recent race meeting all the head stockmen and ringers from the six outstations had come in to the Head Station for the fun, and the manager must have finally found out that it was I who had told a couple of them that they ought to check through their store dockets to make sure that they'd got everything that was booked out to them and debited to their accounts. They had checked and then passed the word around to others of the debited goods they hadn't received.

As I packed all their orders I knew exactly what they'd got, and I'd accidentally discovered the additional entries in the docket book. It was an easy thing to do, because employees were not paid regularly, but by a single cheque when they left or went on holiday. Everything they needed was bought through the store and debited against their wages account. A man who worked for a year would pick up a sizeable cheque and not be likely to go through the twenty or more dockets he might get, along with his cheque. A bit hard to remember, too, what he'd actually bought some months ago, and harder still to prove it. This time, however, the entries I'd noticed were added to the purchases made on the first day the men had arrived in for the races and as they were drawing money to spend at the races they were quite entitled to ask for their dockets.

With the cargo of grog the Wyndham publican had brought to the races there were a few resulting fights, but nothing compared to the row that erupted when those tough stockmen discovered how they were being cheated. I'd served everyone; the additions weren't in my writing; the store was locked at six each evening and only the manager and his newly-appointed assistant had access to the docket book after I'd put it away.

I reminded the manager that he had to give me a week's notice and dear old Peter, the Aboriginal storeman, found me a lad who would, if I lent him my horse, track down the stock-camp and take a letter to Joe.

I only expected a note in reply but, a few nights later, Joe rode in himself. It was against his principles to pull out and leave the camp short-handed, he said, but at the end of this mustering round he'd finish up and we agreed to meet, three months later, at the Negri bush races, over the border in Western Australia. A hour later he was on his way back to the camp.

I didn't get my wages cheque until just before the plane departed; all my dockets were correct but the pay was three weeks short. Joe's cheque, when he got it, was five weeks short. There wasn't anything we could do about it, and we were only two of many, but we heard an interesting sequel a year later. Lord Luke's nephew, fresh from England, presented himself incognito as a station hand and got the same treatment as the rest of us. I would have loved to have been there when he finally revealed himself and sacked the manager out of hand.

It wouldn't be fair to say that the employees on Victoria River Downs didn't enjoy life, or that conditions were very much different from those on the other big Company stations of the North, such as the widespread Vesty Empire. It was still frontier country and, while the Aboriginal population remained for the most part living as fringe dwellers round the homesteads in their own tribal districts, the white and black population was semi-nomadic, taking jobs anywhere from The Gulf and western Queensland, across the Territory and into the Kimberleys. Discounting Darwin, the white population was so sparse that if you hadn't actually met everyone at some time or other, you had at least heard their names or nicknames and some exploit, real or exaggerated, in which they had taken part.

There were few secrets in a land where communication was by means of the transceiver and all telegraphic traffic was public. When Joe and I were married on the spur of the moment at the Katherine race meeting in June 1952, I sent a telegram to my mother in Perth. My flustered parent sent her reply addressed only to "Marie and Joe Northern Territory" and we got it the same day.

We'd both covered a lot of ground since we'd met again at the Negri Races in 1950 and returned to Darwin, where I had been employed as a clerk at Larrakeah Army

Barracks. We hitch-hiked across the continent to Port Augusta and then tossed a coin to see whether we should go east to Sydney to visit Joe's folks or west to Perth to see mine. I won and we caught the train to Perth.

There was a saying that if you ever drank the water of the Victoria you'd always return north again. Actually, I think they also said it about the Ord and all the big rivers of the north. I certainly didn't expect it to apply to me, whose sojourn in the north had been merely in search of copy as a writer, and I'd been away from Perth for less than a year. Why, then, did the city seem such an alien place to me? Why did my old acquaintances seem to be on a wavelength so out of tune with mine?

I took a job as secretary-bookkeeper where the boss had roving hands. Joe, boarding with friends both at Teachers' Training College, was persuaded to sit for the Mature Students' Entrance Exam, which he passed easily, and was enrolled as a trainee teacher. He stuck it out for six weeks and was bored beyond endurance at the prospect of riding herd on little kids in grubby classrooms in place of piker bullocks on northern plains. He threw it up and went contract tree-felling in the bush until he'd earned enough for a quick trip back to Sydney to see his family.

We honestly tried. We'd decided, as the custom was then, to save up to get married and find a settled occupation. A station ringer was always single; no place for a wife when a man's home was his swag, or any hope of marriage until he worked up to a manager's position or at least an overseer. Joe had no ambitions in that line; he'd seen too many managers sacked on the whim of an absentee landlord with relations or friends suddenly aspiring to his job. Nor did I fancy the continual round of entertaining the owner's patronising family and friends during the Dry season that always fell to the lot of the unpaid wife of a manager.

Any place Joe managed, we decided, would be our own. While my married suburban friends were making down payments on quarter-acre blocks we aspired to a couple of thousand square miles in the north. Nobody took us seriously.

In Sydney, Joe was alert to any opportunity. The computer industry was in its early infancy and a big American company was searching for Australian trainees. The boy from the bush, still clad in broad-brimmed hat, stockman-cut trousers and high-heeled elastic-side boots — the only clothes he possessed, — strolled in, sat for a barrage of mathematical tests, was interviewed, photographed, and a fortnight later, advised that he had been chosen as one of the six Australians to go to America to join the vanguard of an exciting new industry. His life was at a crossroads. On one hand, a sure career on a weekly salary equal to a month's wages for a station manager: on the other, a dream.

Joe chose the dream and flew back to Perth, where, a week later, he found a job with Wesfarmers Stock and Station Agents. He was still on track.

In the meantime, I'd got sick of dodging my boss around the office desk and had applied for a job as a reporter on "Hoofs and Horns" magazine in Adelaide. Ten days after Joe got back to Perth, I left for Adelaide. We set ourselves the goal to save 200 pounds each and then we'd get married.

I loved my job. The owner-editor of the only magazine read in the outback was the already-legendary R.M. Williams, founder and owner of the mail-order business supplying saddles, boots, bushman's gear and clothing to the Outback. He sent me with a notebook and camera on trips back to the Territory. When my savings reached 60 pounds I bought a small second-hand motor-bike. Then Christmas-time came around and I had enough saved again for a return fare to Perth. Joe blew what he'd saved on the engagement ring. He'd just been promoted to manager of the Despatch Department and we estimated about six months for both of us to reach the 200 pound goal.

It might have worked too if Joe hadn't been so naive where girls were concerned; after all he'd not had too much contact all those years in the men-only environment of the stock-camps. Jan was the daughter of two doctors and lived in a prestigious home on the Swan River foreshore. She invited him to weekend tennis parties, then to dinner at home. She always seemed to have two tickets her parents couldn't use for some expensive social outing. Each week Joe's letter described these outings and I, who well knew the female wiles, began to bristle. Sure, I was still leading a gay social life but only a group one.

When he wrote about the beach barbecue she'd organised where he'd played his guitar and sung, that did it. I wrote a terse letter saying I could read between the lines and posted back the engagement ring. My Melbourne friend, Pam, had just arrived on a holiday, nursing her own broken engagement. I told RM I was going free-lancing again and Pam and I put the bike on the train and bought tickets for Alice Springs.

At that time the train trip could take anything up to a week. We arrived late one night and went down next morning to collect the bike, which we planned to ride to Darwin. There was Joe leaning against the goods-shed doorway.

He'd told them at work he wanted compassionate leave; they'd refused. "O.K." he said, "I'm leaving!". Reluctantly, they gave him a fortnight. He'd flown to Adelaide, found out my destination from my friends, and flown to Alice Springs to beat the train there.

This should have proved something to me, but it didn't. Pam and I went about our plans and Joe, who hadn't even waited to equip himself with a swag, followed after us. The bike played up and needed parts, so I left it at a garage and Pam and I began to hitch-hike up the Stuart Highway. Sometimes Joe got a ride on the same truck, sometimes not. I clearly remember the truck-driver who put the three of us down at the Daly Waters turn-off. He'd listened to my rude comments to Joe for a couple of hundred miles. "Are you *that* sure you want her, Mate?" he asked Joe.

Pam and I sniffed.

To cut a long story short, in Katherine we met Charlie Schultz from Humbert River, where my sister was governess; we went back with him via V.R.D. Pam's holiday ended and she flew back to Melbourne; and each week Joe sent a telegram to his office in Perth, saying "Back next Monday".

We accompanied the Humbert River contingent back to Katherine for the annual races. We camped on the river bank. Joe didn't have his guitar with him but someone produced one and persuaded him to sing. Next morning, as all the visitors began to pack up their camps, I sensed the change in Joe before he even spoke.

"You can marry me today or not at all!" he said firmly. I knew he meant it. Interested onlookers took his side. "Go on!" they urged. "Okay!" I said.

Someone alerted the Salvation Army padre; Charlie Schultz organised the wedding breakfast at the hotel; I bought the only white dress in the hawker's van, two sizes too big, but altered by the town dressmaker, who also lent us her wedding ring for the ceremony; Joe bought the customary long whites at the store. It was a busy afternoon but the whole township helped. Someone located some roses and made a bouquet; someone else iced two Big Sister fruit cakes to perfection. We were married at the padre's home that evening and the government grader driver with the only film and flashlight in town took our wedding photos for us. The hotel shouted us a room for a wedding present.

Next day we said goodbye to all our well-wishers and got a lift to Darwin to buy a wedding ring. Joe sent a telegram to Wesfarmers: "Back next Monday." Then we hitch-hiked to Alice Springs, picked up the motor-bike and entrained for Perth, via Adelaide.

Joe had been away five weeks and didn't know if he still had a job. All those fares and expenses had added up; we landed in Perth with twelve pounds between us, ten of which Joe owed as back rent on his flat, which he paid the landlady upstairs as soon as we'd unpacked.

A knock sounded on our door. When Joe opened it, my mother and stepfather stood there, arms loaded with steaming dishes and parcels and a bottle of champagne.

"We've come for tea," announced Ferg. "Brought it with us, because we didn't think you'd have time to shop after the train got in!".

"I suppose you've found out she can't cook!" said Mum, hugging Joe first and me second.

"No worries, I can!" laughed Joe.

I imagined Mum saying to Ferg, "All that interstate gallivanting-around must have cost them a pretty penny!" as she packed up a good week's stores for us and cooked casseroles, cakes and pies. With Joe's fares to work for a week to come out of the remaining two pounds, her culinary contributions were more welcome than she could ever have realised.

> *. . . Aged just fifteen … he put together a couple of blankets and a change of clothes, a writing pad and his beloved guitar and headed north for Queensland and the cattle country.*

Joe, ready and willing to fight his weight in wildcats.

> *... I could change a wheel, clean the battery terminals, rope the exhaust pipe back on with a bit of fencing wire and dig out of a bog but I was always a bit shaky about what went on under the bonnet.*

Marie - The 'Tough Girl
of the Territory', 1951.

Bogged West of
Lewis Range.

Chapter 3

Hooker Creek Aboriginal Settlement in the Desert.

Joe did still have a job. The manager at Head Office said that anyone with the initiative and dedication he'd shown in pursuing his goal was just the sort of employee the firm needed and added that he thought Joe had a big future with them. At 24 he was the youngest Department-manager by a score of years and he'd already initiated a number of cost-saving measures.

My plans for finding a job were soon thwarted. I was pregnant and in those days nobody, but nobody, employed pregnant women; the government, and many private firms too, wouldn't even employ childless married women. Besides, I had my hands full learning to cook. I'd avoided what they called domestic arts as a school subject and had never been in any situation which required me to cook for myself. It was Joe who lifted the lid off the kettle and showed me the little bubbles whose presence was necessary before I poured the water into the teapot. Sometimes we had visitors for tea but it had to be on a Saturday so Joe could cook the meal.

In time he taught me to cook. I also picked up a few hints from the two Chinese students from Singapore with whom we shared our downstairs kitchen. O So Lo (who wisely called himself Henry) – was in his final year before returning home to an arranged marriage, a real personality boy with a wild sense of humour, who whooped it up with the girls as if they were going out of fashion. For him I suppose they were. Stephen was a Christian, a steady quiet medical student who later married an Australian girl and stayed in Australia.

My versatile husband continued to amaze me; every day I learned something new about him. There didn't seem to be anything he couldn't do, either with his hands or his head, and do it better than most. I, the independent feminist, soon became his hero-worshipper. I suppose I still am. The only thing he couldn't do was get rid of Jan, who still phoned him at work, demanded he meet her at lunch-time and continued to invite him to dinner at her home. He was genuinely worried that she might carry out some of her passionate threats. I couldn't understand why her parents condoned the invitations to her home, which now, of course, had to include me. Finally I agreed to go and a car was sent to pick us up. Joe had long insisted that he had never encouraged her in the least and I, knowing his strict and almost Quakerish moral code (who better?), firmly believed him.

I honestly thought I was dreaming. Was I expected to be so over-awed, so piqued, that I was going to toss Joe back into the ring the way I'd done during our engagement? I'd been vaguely sorry for Jan, but no longer. She had seated Joe beside her at the long table and I was yards away up the other end with the rest of the family between us. A decorous maid began serving an exotic meal. At heart I'm a thwarted actress and that spontaneous exhibition was probably my finest hour. If Jan planned on showing Joe the vast social and pecuniary difference between what he had, and what he could have, I was more than prepared to help her. I tucked the linen napkin into the neck of my dress and reached right across Mum for a dinner bun, cheerily asking, "You mob say grace? Two, four, six, eight, bog in, don't wait!"

I must say the parents were very good, after their first fleeting expressions, at playing the perfect hosts, though it took Joe a while longer to catch on, distracted as he was by Jan's low-voiced conversation beside him. Deliberately using the dinner cutlery for the entrée, I shouted down the table to him, "Sure beats salt beef and spuds, Hon, don't it! Bit light on though!". I saw Joe turning red and Jan staring at me. We'd never met before but she immediately knew exactly what I was about. I turned to Dad and said cheerily, "You oughta see the feed Joe can put away when he tries. Beef, kangaroo, goanna — you name it, he'll scoff it. Gotta have onions though. You're a doctor, you'd know. Nothing like onions to stop you getting Barcoo Rot!"

That reminded me of a story about a man who claimed that if you got stuck out in the scrub for weeks without any onions you could cure the sores with a poultice of piss and gumleaves. For a moment I thought Mum was going to choke, but she recovered herself quickly. With each course I regaled them with the most bizarre of the yarns I'd heard outback. I particularly liked the one about the ringer stuck on a station out from Derby during the Wet whose mate died suddenly. He reported it to the doctor on the transceiver and was told he'd have to bring the body in for an autopsy. The doctor was a relieving doctor from Perth. He didn't know that the only means of transport during the Wet was horse and pack-horse. When the ringer dutifully turned up ten days later with the remains slung over a pack-horse, he announced, " 'ere y'are, Doc! Most of 'im's 'ere, but after a coupla days on the track 'e got a bit on the nose so I 'ad to gut 'im!"

Mum excused herself hurriedly and went out to the kitchen. When she got back Joe was telling the yarn about the Sister at the Derby Hospital who had a long-running feud with a local character who drove the night-cart, but she never seemed to win a round. She thought her moment had come one early morning on the way to work when she came across the truck slewed across the road, the contents of the cans splattered far and wide and the driver scratching his head perplexedly.

"Ho!" she said, with gleaming eyes. "I see you've had an accident, Jack!"

"Why, no, Sister," he replied. "Not at all. You see, I'm stock-taking, and would you believe it, I'm *one shit short!*"

I'll swear Dad was going to laugh but he caught the stern glance from Mum. My table manners deteriorated further and so did my stories.

They left Jan to have coffee alone with us in the lounge. She was very quiet and there was no offer of the car to take us home. I was a bit apprehensive when we started off on the mile-long hike to the bus stop. I glanced sideways at Joe and saw his lips begin to twitch. We giggled most of the way to the bus stop and then Joe said seriously, "I reckon, Honey, it's time we made tracks back up north."

I thought so too. Privately, I was by no means as confident as I outwardly seemed. Joe was so completely unsophisticated as regards the opposite sex and his uncomplicated sincerity was enough in itself to appeal to a girl, along with his old-fashioned good manners. When he played his guitar and gazed at a girl anything could happen. I ought to know. (He'd better *not* have sung *my* song to anyone else!) I knew better than to rubbish Jan; in all fairness, he would have remonstrated with me. I wondered which of the many love-songs in his repertoire was *her* song. It was definitely time to move on before Jan got her second wind.

It's funny how Fate takes a hand. I had mentioned to a senior editor I'd known from my cadetship days that Joe and I eventually wanted to go north again and this man happened to be a personal friend of Paul Hasluck, who was then the Minister for Territories. Through him, Mr Hasluck (later Sir Paul) learned of our aspirations and, only a week or so after our abortive dinner engagement, Joe got a phone call from the Minister's secretary asking him to come for an interview.

Unlike many politicians, Mr. Hasluck really did understand the problems of the outback and he saw, too, the future potential of the Northern Territory. "It's so sparsely populated yet," he said to Joe, "and we need young enthusiastic people like yourselves there who plan to stay. "Where would you like a job?"

Until then, taking a government job had never occurred to either of us. In fact, throughout the bush, government employees were generally referred to as "two-year tourists". With their high salaries, supplied accommodation, free holiday fares and hearts and minds still firmly fixed in the south, we'd had little in common with those we had met, with the exception of the policemen who were, whatever the reason for one's personal interactions, definitely considered as real Territorians.

There was a job vacancy with CSIRO in Katherine and another in the Native Affairs Department at Hooker Creek, south of Wave Hill, where a small cattle herd was running.

"Well," said Joe, "I guess I know more about raising cattle and working with Aborigines than I do about agriculture, so I'll opt for Hooker Creek. But I'll need to give at least a month's notice at work to finish up some projects there."

So it was agreed. We were delighted; my mother was horrified; our city friends thought we were mad; Joe's employers were more than reluctant to see him go.

I had my appendix removed in a sudden operation and remarked to the doctor that it was lucky it happened before I went bush again. My big mouth! He'd heard all about the isolation of the wet season in the north. The baby was due first week of March.

"How far from Darwin is this place you're going to?" he asked.

"Er — about 700 miles."

"White population?"

"Er — two men, one woman."

"One man and the woman being you and your husband, I presume," he said dryly. "I hope this baby has more brains than you appear to have, my dear."

It was useless for me to argue that I knew plenty of women in the bush in like circumstances. Everyone ganged up on me, and when Joe threw his weight in against me too, I was forced to give in and stay behind.

In early December Joe flew alone to Darwin, just in time for the wedding of my sister Helen to Ray West, whom she'd met at the Katherine races six months earlier. After a few days at the head office of the Native Affairs Department, Joe flew to Hooker Creek, roughly a hundred miles south of Wave Hill, where the Superintendent and his wife were awaiting his arrival, prior to going on leave before taking up a new appointment.

On January 3rd, 1953, Joe signed the handover papers and was officially in charge of the settlement of two hundred tribalised Aborigines, very few of whom even spoke pidgin English. Despite the preceding week of heavy rain, during which Joe spent some time each day sketching his new surroundings, the regular mail plane was able to land two days later. Joe's diary entry reads:

Well the plane landed O.K. much to the relief of John and Helen. However, the relief was short-lived because on taking off on the heavy 'drome the wind changed suddenly and the down-draft pushed the plane nose down into the scrub. She settled back on her tail but was badly damaged. Very luckily no one was hurt and this was certainly a relief to me. I think I broke records getting up the drome and all sorts of horrible things were going through my mind. Fred Ogden (the pilot) is very worried, I think, although trying not to show it. He says the plane is a write-off.

There were another five days of intermittent rain before a rescue plane could land to pick up the pilot and the travellers, leaving Joe the sole white inhabitant for a week until the next fortnightly mail plane arrived, bringing him an assistant, Paul Ingram.

Paul was a character with a typical outback sense of humour who, when he saw Joe's enthusiastic plans for the settlement, announced that he was really Spanish and his name was Manuel — Manuel Labor.

The settlement consisted of a fenced-in Superintendent's house, built by the first superintendent, Bill Grimster, about five years previously. It consisted of one large bedroom, a large lounge-dining room, and a kitchen and pantry, with a wide fly-wired verandah all round with one corner partitioned off for a bathroom. It was built of cement-brick, cement floors and a small building adjacent to the kitchen end housed the toilet and the laundry, complete with a copper and pair of cement tubs.

A second building, a corrugated-iron ex-army Sydney Williams hut, faced the homestead 150 yards away. This was divided into two sections, one for the store and the other for the hospital. A two-door metal cupboard housed all the hospital equipment and the half-dozen beds were also ex-army — cyclone wire folding stretchers complete with thin horsehair mattresses, the same as supplied as part of the house furniture.

Making the third side of the square was another corrugated-iron hut, roofed but dirt-floored and with only half the walls erected. This was supposed to be the residence of the Assistant Superintendent.

The square was completed by the workshop and fuel dump, again the ubiquitous corrugated-iron, with a saddle shed added like an afterthought to the far end. An airstrip and a bore equipped with windmill and stocktank made up the settlement.

None of these edifices had been erected by tradesmen on contract. The two previous Superintendents, working with what the government had supplied of hastily and cheaply bought ex-army building materials and whatever they could scrounge themselves, with any amount of cheap Aboriginal labour (though none of it in the least experienced), had done a very creditable job under the circumstances.

The blacks themselves still lived as they had done for centuries. A half a mile away, a collection of wurlies and humpies, known as 'The Camp', housed the tribe and their multitude of dogs. There had been delivered, however, materials for the erection of twenty corrugated iron huts — one to a family— and this was to be Joe's first major project.

Each day began with hospital parade, normally a couple of hours. This meant handing out "coughin' medcin" and "rubbim' medcin", goats' milk to children and expectant mothers; and, on a fairly regular basis, treating spear and nulla-nulla wounds. The Walbiri were a fiercely proud and independent people and settled their differences as they had always done. Their fights often occurred at night and an uproar of shouting and barking of dogs invariably alerted the whites to stand by the flying Doctor Medical Kit ready to patch up the wounded.

In mid-January, when Joe reported his ministrations after one particularly bloody fight to the Flying Doctor in Darwin, the incident was reported in all the southern papers and the sub-editors changed the appellation of Superintendent to Doctor A. J. Mahood. After all, who else but a doctor could extract spears right through limbs, sew up deep knife wounds, set broken arms, and treat patients unconscious after a belt on the head with a nulla-nulla? Little did the Southerners really know of life in the outback. Joe was also keeping alive a two and a half pound premature baby rejected by a mother suffering post-natal depression, and doing his best to persuade hoary old warriors with headaches to take an aspro instead of binding the offending ache tightly with hairstring.

The Wet season stores due the previous November had not arrived and the store was almost empty, so Joe was obliged to kill a bullock at least once a week and further ration the dwindling flour, sugar and tea supplies. The Aborigines, still fully tribalised, had lost none of their hunting skills and those not on strength as "working men" went bush with spears and boomerangs to augment their rations, though it was almost desert country and could never have supported anywhere near the number of people at the settlement.

Joe, when he flew in to the job, had taken with him some packets of vegetable seeds with the idea of starting a private garden. These he now used to start a settlement garden down near the stocktank, where it could be watered easily. His garden needed constant supervision and most of the work he had to do himself because none of his charges had the least idea of agriculture.

His essay into town-planning on the slight rise near the camp was moderately successful. He and Paul, with the help of the working men, erected the huts and duly allocated them, taking care to consider the rigid skin-system divisions of the tribe and the mother-in-law skin taboo so that no woman lived in a hut too near those of the men of the tribal skin she was forbidden to "look at." With his alert and curious mind, Joe had already observed much of the Aboriginal social organisation in various tribes, but it was at Hooker Creek, where no hint of tribal breakdown had yet occurred, that he was able to learn all the ramifications. He knew about the matrilineal descent of the skin system and here, where there were a few "little-bit straight" marriages as well as "straight" marriages, he had to be careful with his allocation of the huts.

Privately, he was not sure whether housing the people in European-style, and therefore quite costly, huts was altogether a good idea, because, in the event of a death in one of the huts, tribal law decreed that the hut and all the deceased's belongings must be instantly burnt and the name of the dead person never again mentioned. This applied to animals as well, and the reason was to keep at bay the restless spirits of the dead which, jealous of the still living, would make mischief for anyone who inadvertently spoke their names or lingered near the place of death.

Joe's priorities were a little different to those of Head Office. He thought toilets and shower rooms, of which none had hitherto existed, were more important than the huts, one of which he cannibalised in the interests of hygiene.

He was later stunned to find out that, unlike the average white station inhabitant or the semi-nomadic ringers who worked side by side with Aborigines in the outback, none of the administrative staff of the Native Affairs Department, who made and applied the policies, seemed to have the remotest idea of the social system or the tribal laws of those over whom they played God. Appointed from Canberra, whose ivory-tower academics didn't even have a clue of what made the Territory whites tick, let alone the Territory blacks, these administrators were guilty of some appalling blunders.

Little wonder then, that, with few exceptions, the white staff on the settlements saw themselves, together with the Aborigines, as US against THEM at Head Office, where they apparently thought that "skin" meant what colour you were and were unaware that there were only eight possible skin-names in a tribe (which pass for surnames), so that visiting patrol officers and clerks all wrote down their own versions of a name as they heard it from its owner.

A simple instance. Joe employed among his working men one Tommy Djungarai, who, in official correspondence was referred to as Tommy Chungaree, Joongery, Tungary, Toongaree and Djangri and hence became six different individuals. This lad had been to Darwin and called in at the office there a few times. Two or three of the other workers also had extra names.

Joe's efforts to straighten this out met with singular lack of success and a belligerent determination that this fellow who was trying to rock the boat needed to be put in his place. How much easier to shove the necessary thumbmark on the extra pay cheques (any thumb mark would do, nobody ever checked) and pocket the cheques so the paperwork in Head Office came out right.

But Joe had integrity and, young as he was, his name was already respected throughout the inland Territory for complete honesty and straight dealing by white and black alike, even by hearsay from folks who had never met him. He *had* to send the cheques back, and he *had* to say why. This, on top of his other requests like "Please when can we expect the loading of stores to feed population of 200, now four months overdue?" did nothing to endear him to the bureaucrats.

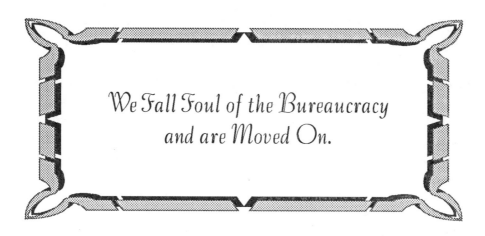

Chapter 4

We Fall Foul of the Bureaucracy and are Moved On.

On the 2nd of April I arrived at Hooker Creek with baby Kimmy, just over four weeks old. It had been a long journey from Perth — first day, fly to Derby; second day, fly to Halls Creek; third day fly to Wave Hill; and there Joe met his wife and new daughter in the battered old utility which took eight hours on the following day to cover the hundred miles of desert track south to Hooker Creek. Oh, the joy of reunion and all the idealistic plans and the challenge of it all. Anything Joe ever did he did with all his heart and running Hooker Creek was no exception; I could not but be inspired and caught up in his enthusiasm too.

Paul's wife, Tessa, and their toddler, Peter, had arrived a fortnight earlier and Paul had patched up a basic dwelling for them in the half-finished Sydney Williams hut. We became instant friends that first evening. I inspected the homestead with glee; this was to be Joe's and my first real home together, and I had lots of half-formed ideas of how I was going to run it, none of which came to fruition, because I found out very shortly that the boss of that homestead was indisputably Nancy Nabaldjari — fat, black, darling Nancy Nabaldjari— who had housekept efficiently for Joe and Paul and wasn't about to resign her position to any little skinny whitefella missus.

Nor did she think much of my child-rearing abilities. She took over my "proper prettyfella picaninny", renamed her Winingali, fashioned her a coolamon from Leichhardt pine, gave her a dreaming — Binta-binta (Butterfly) — and allowed me to feed and only sometimes bath the baby under strict supervision. Joe was "Mulleka" (Boss) but even he had to toe the line to suit Nancy's routine to a certain extent. While she grumbled about his untidiness and picked up after him, she gave me to understand in her inimitable pidgin that all men were "no-good buggers" in some respect and it was a fact of life that women just had to put up with it and, all things considered, our Mulleka was a cut above the rest. Dear Nancy. In all the trials and tribulations of that year I think it was great-hearted, reliable, bossy Nancy who made it a time that we always remembered fondly, despite its setbacks.

The setbacks were all related to *them* in Darwin. The actual running of the place,— the incidents like hospital emergencies, accidents, equipment breakdowns, mediating in the often-violent fights — all these Joe handled with aplomb, as well as steadily improving the settlement status. The vegetable garden was now flourishing with two gardeners trained to look after it. The cattle herd which roamed free of fences and watered at a bore fifteen miles away was finally brought under control. There were only seven horses on the place and only one young fellow who had had any experience as a ringer. Joe trained half a dozen youths as stockmen. He solved the horse shortage problem by trapping brumbies running on Winneke Creek to our south, and breaking in the likely-looking colts. He repaired the inadequate yards, and mustered and branded up the calves by the bronco-panel method.

Bronco-branding was then the common practice in the Outback where yards were either simple enclosures or non-existent. The bronco-panel consisted of two wooden yard panels with the sturdy centre posts set just close enough together to enable a rope to slip between. This could be built either inside a yard or out on the flat. A ringer mounted the bronco-horse, a large strong horse often with a touch of Clydesdale or Percheron in its breeding, rode into the mob, selected and lassoed the cleanskin calf and dragged it towards the panel. The horseman rode around one end of the panel, flipping the rope through the centre posts as he went. Ringers waiting at the panel then threw leg-ropes on the bucking, bawling calf as the post brought it up short, held it down and ear-marked and branded it and castrated the males. All the while, the bronco-horse stood braced at the front of the panel and held the head-rope taut to the saddle pommel until the job was done and the calf released to run back into the mob.

In all the dust and shouting and bellowing, the ringers had to be agile and constantly alert because a savage kick from a struggling mickie or a slip with the razor-sharp knife could mean an injury sustained far from medical help. Be that as it may, the young fellows seemed to love every minute of it and prowess with the lasso meant regular challenges for the envied job of riding the bronco-horse.

The smaller calves were "scruffed", that is, run down and grabbed by two or three ringers and physically held down while another darted in and applied the brand or knife.

The Hooker Creek lads were enthusiastic learners under Joe's tutelage and soon became skilled enough to take jobs with drovers looking for extra men.

In his rare moments of recreation, Joe played his guitar and sang to an admiring audience of piccaninnies attracted when they heard him singing "Beautiful, beautiful brown-eyes" to his little daughter, or he snatched a moment to sketch a sunset, a ragged dead tree, or the character-filled face of an old warrior. On the rare occasions we went riding together, with Kimmy left in Nancy's able care, we always ended up doing a job, bringing in some cleanskin cattle, checking on an overdue hunting party. It just wasn't in his make-up to do anything purely for pleasure with no purpose behind it, which took me quite some time to understand, sybarite as I was. Joe worked on Hooker Creek just as he would have done if it had been our own place, and I shared his satisfaction at each new improvement.

It's hard to imagine now just how inept our Head Office was. Up till that time I'd believed that the jeering comments from the bush people about Native Affairs might have been exaggerated, but our experiences soon made me realise otherwise.

Joe was responsible for the welfare of about two hundred people and the stores to feed them should have been delivered before the Wet last November. His anxious queries sent on each fortnightly mail plane elicited no response. The pilot of the tiny mailplane brought me two pounds of butter and half a dozen oranges each flight, and we had plenty of goats' milk, meat and some vegetables, but the little flour left in the store was grey and weevilly and the other dry rations almost gone, despite the strict rationing.

Our personal belongings, consigned in Perth in February, had not arrived and as I'd flown in with just a suitcase, we didn't have any baby gear, linen for the beds, crockery or anything like that. We ate from the ex-Army tin plates and Joe made pannikins from empty jam tins and fencing wire. His enquiry about our personal gear elicited first a letter that it had been sent to Hatches Creek by mistake and then, a month later, another letter just stating baldly that they didn't know where it was, end of matter. (Some months later my brother-in-law, by sheer luck, noticed crates addressed to us right at the back of the Bond stores in Darwin, claimed them and sent them to us with a transport driver going to Wave Hill.)

But back to our ration crisis. In early May I picked up a telegram on the transceiver session which ordered Paul and family to proceed at once to Darwin in the settlement utility for transfer to Snake Bay. We were really sad to see them go, but at least Paul could *tell* how serious things were with our food stores now six months overdue. In the meantime we now had no vehicle at all and therefore no way of charging the 12-volt battery to run the transceiver, which was our only means of communication with the outside world, apart from the fortnightly plane service. With so many people on the settlement we had numerous calls in the medical sessions to the Flying Doctor in Wyndham and that transceiver was our lifeline. Joe sent urgent telegrams asking for the return of the vehicle. No reply. There was supposed to have been a truck at Hooker Creek too, but that had also gone to Darwin some time before Joe's appointment, never to be seen again.

We eventually found out, after we left Hooker Creek, why none of Joe's requests, urgent and otherwise, had ever been answered. The Patrol Officer nominally in charge of the settlement requisitions had gone on six months' long service leave. Any letters or telegrams from Hooker Creek were his business to attend to, so the clerk just put them on his desk, week after week, month after month. Honestly!

Our kerosene supply, which serviced our two refrigerators and the pressure lamp, was nearly expended, so we turned out the fridges, salted the meat, and Joe quickly made a Coolgardie cooler for the milk, butter and cooked meat, so we could conserve the kerosene for the light. A lot of his medical ministrations took place at night. I'd already seen him set a broken arm and stitch a knife-split lip while I bathed and bandaged minor wounds at ten o'clock at night. We needed that light.

We also needed soap for ourselves and the tribe. We had plenty of fat, having to kill a beast for meat so often, and caustic, but no resin. Joe substituted bloodwood gum collected from the bush trees and it worked a treat. The soap turned out a rich, reddish colour and it lathered okay but it stained the baby's nappies a mottled pink.

Then out of the blue a truck arrived, and with it, bearing a bottle of rum "in case of snakebite" came anthropologist Merv Meggitt. Ken, the truck driver, gave Joe the mail with a letter from Head Office which rather churlishly directed that Merv was to be entirely self-dependent and not permitted to interrupt the settlement routine in any way. Very obviously, Head Office didn't approve of Merv, which endeared him to us straight away. They had passed Paul and Tessa at Top Springs, so Paul wouldn't have to press our case for urgent supplies when he got to Darwin.

What Paul didn't know, of course, was the composition of the load. There were five tons of petrol, of which we still had plenty, seeing we had been one vehicle down on strength, but there was no kerosene. There were cartons and cartons of clothing. Some sugar, tea and jam, but no flour and no soap. The tribe went wild over the sugar and jam when Joe handed out a ration immediately; Nancy gloated over the lone straw broom to replace the bundles of leaves she'd been using; Joe grabbed the box of torch batteries and wired them together to make up twelve volts to substitute for the now defunct transceiver battery.

We were back in touch with the outside world, though not for long because there were only twenty-four little batteries. I hunted anxiously for matches but there weren't any. Joe and I still had some carefully-rationed tobacco, which we rolled up in little squares of toilet paper after a meal. To light them, if the kitchen fire was out, we had to go outside our gate, shout "Warloo!" and an obliging Aborigine would run up and share his firestick with us. (Nancy brought her own firestick from the camp to light the fire for morning tea at five.)

Merv had his own stores with him, and immediately donated some potatoes and onions, along with the rum, so it was a gala tea that night. His wife, Joan, was to arrive on the next plane and they were expected by Head Office to camp in the scrub. Head Office was a long way away; Paul and Tessa's home was now empty. Enough said.

"Flour, please," squeaked our urgent telegrams until the batteries gave up the ghost. No option left but to borrow some from Wave Hill. Joe mustered the few horses, and set off with a string of packhorses. Merv went along too, although he had never ridden before. It took them nearly four days to reach Wave Hill, and in the meantime the tribe unobtrusively looked after Joan and me. Mulleka, unknown to me, had mentioned to the old warriors that the two young-fella missus were in their care and I noticed that, wherever I went, a lone squatting figure held me under observation. There were no fights; the settlement ran like clockwork.

The Wave Hill manager lent the flour and the Wave Hill policeman offered to drive down with a load, arriving just before Joe and Merv got back with the horses. The tribe lined up with ration bags and broad grins. Lovely! And the policeman lent us a 12-volt battery for the transceiver.

Barely a week later a Parliamentary party, led by Paul Hasluck, called in at Wave Hill, en route to Wyndham to inspect the meatworks. The manager told them with relish about our problems at Hooker Creek and Paul Hasluck, on arrival at Wyndham, telegraphed to Head Office and to us that he would arrive by charter flight for an inspection the next day.

Never a dull moment back at the ranch. The bore had broken down and Joe couldn't pull it without a vehicle. He sent a telegram that we needed the vehicle urgently. No reply.

We were nearly out of water and Joe was just about to put into action his plan for all of us to ride or walk to the stock-water bore fifteen miles away, when Ken arrived with a second load. Joe requisitioned the use of his truck immediately to pull the bore while I inspected the load. Flour at last, more petrol but no kerosene, four enormous cartons of toilet paper and about twenty four-gallon tins of ex-army dried peas. The war had been over for eight years and the peas were feeling their age. I will never forget the expression on Paul Hasluck's face as he scooped up a handful, crumbled them and threw them down in disgust.

My diary entry for that day reads:

Paul Hasluck, Mr Howse and pilot arrived and looked over the place. Joe didn't stop talking all the time, and they left, promising us that something would definitely be done. Just recovering from the excitement and another yell from the tribe "Truck comin'!" Patrol Officer George Holden with our ute at last!. Poor George must have driven non-stop from Darwin when Head Office got their telegram from the Minister, but he was just too late to pull rank and take over the conducted tour from

Joe. George got a ride back to Darwin with Ken, and so did Nancy and I, to visit my sister Helen and show off my beautiful child.

It was really amazing how efficient the personnel at Head Office suddenly became. It was in inverse proportion to Joe's chances of achieving the already-discussed promotion from Acting Superintendent to Superintendent. The head man himself was instructed to visit and he came. We found out later that he said in his report that Joe was young and inexperienced and had let the natives rob the store, so that was why we'd run short of rations.

Our own private food stores had arrived with Ken the day before Mr. Hasluck's visit and while I was away in Darwin a medical team arrived for an official survey — a doctor, nurse, two dentists and a patrol officer. Joe billeted them all and co-opted Joan and Merv to help cook for them. When I got back the only items left from my private order were four bottles of Worcestershire sauce — not even one spud or onion. The visitors had eaten the lot.

During his visit the head man had mentioned travelling and meal allowances for government employees, something we'd never heard of. I jumped in and mentioned how much out of pocket we were from the medical visit and he asked Joe for details and said all the visitors should either have fed themselves or paid Joe for their meals as they had a generous allowance to do so. He said he would contact the visitors and claim for us. I guess we *were* young and inexperienced. That man sent Paul Hasluck a bill in our name for the three lunches I'd provided and the Minister sent us a cheque. No one else ever did and nor did the head man pay for the meals for three I'd provided during HIS stay. I could have forgiven him everything else, but sending Paul Hasluck a bill was the last straw. Left to himself, Joe would never have made any claim from anyone; it was me and my big mouth and feeling snaky about coming home to a bare pantry that had provided the opportunity for such spiteful behaviour.

The next move was a telegram to say that a Superintendent had been appointed and Joe was demoted to Assistant and we must move out of the homestead into the other building.

When the Superintendent did arrive on the next plane, he asked me if we would stay in the homestead and board him for some months until his wife arrived. We agreed. He was a quiet middle-aged man and at first I found him very pleasant. I was pleased that Joe would be relieved of some of the workload he'd been carrying single-handed. (Ten years later a staff of forty lived on Hooker Creek with the same number of Aborigines.) But after about a fortnight it was evident that the new man didn't plan on any physical work - I rather doubt he knew anything about killing, butchering, water reticulation, medical chores, mustering cattle or any of the daily chores of bush life; he'd come from somewhere down south. He referred to the tribe as "the niggers" or "the coons" and sat most of the day over the office files.

After their first few attempts, directed by Joe, to go to the boss for their daily instructions (which Joe had already discussed to help him take over), the working men soon assessed the new man's capabilities. The situation deteriorated rapidly. I felt sorry for the poor man because he was so far out of his depth it wasn't funny. The tribe wasn't so compassionate. They sensed his dislike and fear of them and ignored him completely, coming directly to Joe for all their needs. They still called Joe Mulleka and the new boss was "that nother man". It was embarrassing for us but in the end Joe went on as before with the actual work and the boss did the office work (which had taken Joe three or four hours once a fortnight on the day before mail day.)

Merv and Joan had left on an extended visit to Limbunya, so Joe was able to lay an antbed floor, and build an enclosed verandah and a better bathroom for our proposed dwelling. When he'd finished we decided to shift over, so the new boss wouldn't be so exposed to the tribe's lack of respect when anyone came to the house asking for Joe and acting as if the other man wasn't even there.

Joe was a lot more mature than I was. He went to the homestead early every morning and told the Superintendent what he planned to do for the day by saying, "Would you like me to get a killer today?" "If you've nothing else planned, I thought I might clean out the store", and so on. Joe told me that it was important that two men working together in such an isolated situation should get on together, but I heartily resented the man who had the job which should have been Joe's and couldn't handle it, and I didn't bother to hide my feelings. Joe and Paul had worked together like clockwork.

When we shifted to the other house, naturally Nancy came too. The new boss called me after a couple of days and told me to "send that fat nigger woman back to the homestead to work there". He didn't ask; he told me. I spluttered and then said "You tell her". I knew very well how Nancy would react; nothing would prise her away from her little prettyfella piccaninny belonga Nancy. He did tell her and she said nothing but sent another quite competent girl up from the camp the next morning to the homestead and turned up as usual at our place.

Joe added a pantry to our kitchen and reticulated water to the newly planted garden and orange trees in our yard and little Kimmy crowed with delight whenever a black face bent over her and cringed in fright when it was a strange white one. Some real furniture arrived for us — a table, chairs, bed, dressing table and wardrobe. It was cheap stuff and very obviously second-hand, but a lot better than the wire stretchers and wooden boxes we'd been using.

Most of my time was taken up with the school I'd earlier started for the piccaninnies. Joe had partitioned off a section of the hospital, built a bathroom and cut forty-four gallon drums lengthways for tubs, made me a long table and forms and a blackboard, and I had two ladies from the camp to bathe the kids in the morning and issue school clothes for the day, and reclaim and wash them in the afternoon. (I couldn't let them wear them home to the camp because the Walbiri were inveterate gamblers and would stake anything in their card games, including their kids' clothes.)

The Director of Education in Darwin, Lyle Newby, had supplied me with chalk, pads, books and equipment, although he couldn't pay me anything because I wasn't a trained teacher and we weren't an official school. That didn't matter; I loved the job and the children's progress was ample recompense. I wrote a letter to the magazine in Perth, for which I'd once run the Children's Pages, and in return for descriptions of my pupils and their doings, the children's editor started a fund to buy Christmas toys for the piccaninnies, none of whom had ever seen a whitefella toy before. Life settled again into a steady routine and Joe and I were both too busy to register any undercurrents.

My diary entries for three days in late November best tell the rest of the story.

Wednesday 18th. Bolt from the blue. Wire from Darwin to say Mac and the Administrator coming down and we are transferred to Delissaville. It is a horrible shock, especially at this time of year. Got in touch with Top Springs and cancelled six months' stores order. I will have to close the school down and we will have to sacrifice our garden. Have to be packed by Friday, so we will have to leave ever so much behind.

Thursday 19th. Plane arrived today, plus Mr. Newby, who is very upset over closing the school. We were astounded to learn the reason for the move. J. wrote up and asked them to shift Joe as he could not stand the atmosphere because he sensed that we all thought he was incompetent. He's damn right there, the crawling little swine. Mac was a bit staggered to learn that we knew nothing about it. J. tried to get out of paying the board he owed me - said he'd send a cheque, but I said I'd take it now and save him the stamp.

Friday 20th. I left for Darwin in the plane this morning. Nancy is leaving to go to Wave Hill, won't stop after I go, dear old thing. Joe is to come by road with our luggage. I howled when I left, as it had been a wonderful year all round and I hated leaving. Chloe refused to work any longer for J. yesterday, and I was touched by the blacks' feelings for us. I consider them real friends, far better than some whites I know!

It had been a wonderful year too, despite all the administrative mix-ups. There'd been the Limbunya race meeting, where Merv and Joan were godparents when the AIM padre christened Kimmy, and the Hooker Creek contingent had more than excelled in the races. Joe won the Bracelet, the Wild Donkey Ride and the Men's Footrace and Nancy, despite her protestations when I urged her to enter that "them skinnyfella lubra allasame racehorse", came third in the Lubras' race, also won by a Hooker Creek girl.

There'd been the corroborees, both playabout and serious, that Joe and I had been invited to witness (always dependent on the sexual taboos), and Nancy's daily confidences about the illicit love affairs of the camp which added a human side to the strict tribal law, where death or ritual spearing was the punishment for transgressions. (It said a lot for the power of love when some young man would risk his life for secret meetings with a young girl who might be the third or fourth wife of some jealous old warrior.)

There'd been the potentially dangerous incidents we'd experienced too, when Romeo went murder-mad and speared three men during the night. Joe and Paul had to catch him next morning and hold him, raving mad, for five days before Head Office got around to sending a policeman and nurse, complete with tranquillisers, handcuffs and strait-jacket, to take him away in a plane. Head Office hadn't seen fit to advise us of his past, where he'd twice killed in similar episodes at Yuendemu and they were actually going to send him back to Hooker Creek after he recovered in Darwin. Luckily for us, he went violently crazy again the very day before he was due to return and was sent to an Adelaide asylum instead.

When the Health Department sent us the measured doses to administer to the whole tribe for hookworm treatment, they forget to tell us about the side effects. Joe dosed half the tribe one day; the second half planned for the next day. Within a couple of hours they were paralytic drunk, staggering around, hiccoughing and passing out on the flat. The undosed half of the tribe helped Joe and Merv pick up the bodies and lay them in the shade in the big shed, and Nancy stripped off naked and stole Kimmy and went bush. I had to track her up and wait till she passed out before I could get my baby back.

How lucky we were that they were all happy drunks. "Silent" Djambijimba, a complete introvert, staggered up to Joe, looped an arm round his shoulders, greeted him effusively and gave him back the pocket-knife he'd pinched months before; all taboos forgotten, males chatted up forbidden females; the first-dosed hospital assistant made a pass at Merv. It was a very hectic few hours before they had all passed out and the first-dosed were beginning to come round with a hangover, and very frightening for us, not knowing at first whether they actually *would* recover when they collapsed.

Naturally, there was no second dosage, but a good percentage of the tribe turned up for hospital parade next morning and a warrior spokesman cheerily advised Joe that "this mob wantim more wittekee (whisky), Mulleka!". I wonder if that incident was one of the reasons the tribe always reckoned Joe was such a good bloke! What a memorable party he'd thrown! The irony was that the results of the tests done by the medical team during my absence in Darwin were later found to be wrong and nobody had hookworm at all.

What a lot we'd learned in that year about human relationships, both black and white, and how proud I was of my husband who had demonstrated that he could calmly handle any situation, no matter how unexpected or physically daunting.

For all that the Aboriginal women seemed to be mere chattels of their menfolk — young girls handed over to old men in a polygamous society, over-worked, beaten if the old fella was cranky — the women were a much more cheerful lot than the men ever seemed to be. In the camp of her husband, a girl might decorously submit to her head being shaved bald by a co-wife, and the two of them sit meekly all afternoon rubbing the strands of hair on their thighs to make hair-string for the old man, but once it was all girls together there were shouts of laughter and high spirits and often the hilarity was the result of a girl mimicking her old man or his friends. They were absolutely wonderful mimics, as I had occasion to know after Nancy and I returned from Darwin and I accidentally happened on her demonstrating to a select group of her friends how I had comported myself in the capital.

As a result of that visit I was to learn more of what gave some of the zest to their lives and made the brown eyes dance when their menfolk weren't looking. Joe had asked Nancy what she thought of my beautiful sister and when Nancy gave him a most enthusiastic description Joe told her that, seeing Helen was straight skin for him, he might take her for a second wife. Nancy's brows darkened and she told him he'd be out of luck because Helen had a husband "too much jealousin' bugger!" But she was clearly worried; she'd seen a lot in her lifetime in the bush and didn't have too high an opinion of the male sex, black or white.

For days she was constantly re-assuring me that Ray wouldn't give Helen up without a fight, and I should have told Joe to tell her that he'd changed his mind but, even then, I doubt whether it would have altered Nancy's intention. She decided that it was high time he was taught a lesson; I must make *him* jealous and afraid he might lose *me*.

The first I knew of it was when I was invited to a women's corroboree one moonlit night. A laughing group of girls met me at the house and escorted me to their special dancing ground, where I was assured no man would dare to venture. Nancy was obviously in charge. This, I was told, was special magic, and all for me "to sing 'im up 'nother man for Missus". They sat me in the middle of a circle, tapped a rhythm with their carved sticks, chanted, and performed some particularly erotic dances for the best part of an hour. I was "sister belong them" and obviously there were times we women had to stick together.

Three days later two young ringers from Limbunya station visited Hooker Creek for a few days and the tall, really handsome one immediately took a shine to me. Nancy dug me in the ribs in the kitchen and smirked suggestively. What sort of mental waves those girls were sending to their victim I'll never know, but every female I saw over those few days would give me a conspiratorial glance as Lucky's behaviour became more and more embarrassing. I gave him absolutely no encouragement and when he cornered me and told me he intended to take me away from Joe sooner or later, I could hardly wait to find Nancy to tell her she'd better undo the spell quick smart.

"Can't do 'im, Missus," she responded cheerfully, "you gottim dat fella allatime now."

At the races a month later I did have him too. It didn't work quite the way Nancy intended though, because when I appealed to Joe not to leave my side he took it as a compliment to himself that he had such a popular little wife, and didn't show the slightest sign of jealousy. Nancy said it was my fault because I hadn't played my part properly and anyway it must have worked a bit because she never heard any more mention of Helen from Joe. She assured me that the girls could "sing 'im up" any specified man for any girl but seeing I didn't have a particular man in view they had worked hard to conjure up a tall, handsome, virile one for me.

No wonder the women were comparatively so cheerful, with a weapon like that in their control — we could have made a fortune with it if we could have exported it to the whitefella world! But of course the whitefella world is too sophisticated to believe in blackfella magic. I do, though. I have, over the years, seen too many examples of events which cannot be explained by ordinary whitefella logic.

Chapter 5

A Wet Season in the Jungle.

The November wet season loading did arrive, not surprisingly, that year and Joe returned to Darwin with the truck and our personal gear including the cot and high chair he'd made for Kimmy. We'd only had our gear a couple of months after its loss in the Bond stores and I think it must have been fated. The truck was caught in an early monsoon storm, bogged in Bull Swamp and finally bogged again at Redbank, where rising floodwaters went right over it and Joe and the driver had to foot-walk the last 25 miles in to Katherine. It was a week before they could get the truck out and continue on to Darwin with the sodden and mouldy mess still crated on the back. We were able to rescue the wedding present crockery, cutlery and a few other items and the government insurance inspector gave us one hundred pounds in compensation for the rest. We blew the lot, not on replacements, but on a complete kit of good tools for Joe and a portable typewriter for me, both of which were to play quite a big part in our future occupations.

Delissaville, across the bay from Darwin and up the tidal Delissa Creek, was as far removed from Hooker Creek as it was possible to be. From desert to jungle in one move. Our dwelling here was the original donga erected by the first settler in the region. It was a pitted cement slab with a roof over it, one central room with doorways but no doors, and walls of laced bamboo to partly enclose a verandah. One end was partitioned off with sheets of corrugated iron, behind which a pipe with a shower rose stuck out of the wall and a grubby hand basin lurked, and at the other end was a small free-standing stove, a kerosene fridge, a sink and a table and two chairs — the kitchen.

A set of laundry tubs was positioned against the outside wall and an old copper stood against a sagging netting fence, just far enough away from the dwelling not to burn it down when the fire beneath the copper was lit. Joe's first idea that he could use his new tools to make the place habitable was quickly discarded. "One hammer blow, Hon," he said, "and the whole darn place will fall down about our ears!"

We got seventeen inches of rain the first night we were there. The water flowed in through the front doorway and out the back one. We put the baby's cot on the table, covered with Joe's birkmyre swag cover, and waded round up to our waists in water with our suitcases of clothes on our heads. In the torchlight I could see frogs and centipedes and God knows what else swirling by. There were crocodiles in Delissa Creek and by my reckoning Delissa Creek was in our dwelling. Something brushed against my arm and I yelped. It was only a floating chair.

"Buck up, Hon!" shouted Joe above the pounding din, grabbing at the table so it didn't float away too with its precious burden.

In the morning we found that the only thing that actually had floated away was the wooden dunny at the bottom of our yard. Joe was an old hand at making bush dunnies so that was no real loss.

Tas Festing was the Delissaville Superintendent, one of the original old hands, who had started Borroloola Settlement from scratch. He and his wife Pat were great people. After three weeks Tas asked Joe curiously, "What on earth did you do to stir up the Head Office boys? I've been instructed to watch you like a hawk and to write a report about you. Sounds like they want an excuse to sack you." Joe told him briefly. Tas laughed. "My report says that I can't stop you working and you're right on the ball and worth three of any other blokes I've ever worked with."

At this time Bill Harney was living under the big banyan tree on the beach at Two-Fella Creek, about fifteen miles north of Delissaville. Bill's prestige as an author and poet was by now well-known down south and abroad but it had changed his lifestyle not one whit. Bill had sampled every experience and tried every type of work that ever existed in the bush. He knew the Aborigines and had a bond with them as few other white men before or since.

He was known and trusted by Aborigines throughout the Territory. They called him Bilarni and shared every facet of their lives with him. In the course of his varied career he had been a patrol officer in the Native Affairs Department just five years before Joe joined up and had, after he left the job, written a book entitled "Naked Affairs". I don't know the exact truth of the matter but the book never saw the light of day. It was about Bill's experiences as a patrol officer "and it's dynamite!" I was told by one who had read the manuscript.

Who suppressed it, and how, I don't know, and Bill, whom we were to get to know quite well in later days, would never elaborate. All I know is if Joe was out of favour at Head Office and the mention of his name brought a frown to administrative brows, then the mention of Bill's caused the nearest thing to an apoplectic stroke.

Our donga stood fast even during the cyclone, but we sheltered in the hospital that night. I had earlier gone there to help Pat if needed when an Aboriginal woman with a history of miscarriages had gone into early labour. Tas and Joe wanted to take her across to Darwin in the "Amity", the settlement launch, but Pat and I soon vetoed that idea. The girl was about six months pregnant and it was obvious that the baby was coming soon anyway. She had enough problems without being out on the bay in a cyclone as well. The tiny boy she delivered was little more than a pound in weight and despite our efforts only lived for an hour and a half.

Though we only stayed there four months, we enjoyed our sojourn at Delissaville. The Aborigines were very different from the light-skinned, fair-haired desert people, much darker with black curly hair, finer-featured, and many with a definite Islander or Malayan heritage. They were fantastic dancers, could have graced any ballet, and, almost without exception, could sing wonderfully. At dusk most of the young people would gather under a huge milkwood tree and sing to the beat of drums and didgeridoo the latest hit songs and the haunting Islander melodies.

Someone heard Joe playing his guitar one evening and he was immediately invited to join them. How they could harmonise! In the velvety tropic night, with only a campfire for lighting, the magnificent voices of Joe's chosen quartet floated into the cathedral of the sky in a harmony and rhythm I've never heard equalled since. Except when it rained, "choir practice" was our standard evening entertainment.

The tribe was the Larrakeah, a much more sophisticated people than the desert Walbiri because of their close proximity to Darwin, and while all spoke pidgin, — the lingua franca of the north which gave the people of the various tribes a common second language, — many of the Larrakeah spoke excellent English as well.

I thought, and still do, that pidgin was one of the most colourful languages I had ever heard, spoken with the lilting cadences of the Aboriginal languages and rich with graphic similes. I remember Nancy, one washday, dangling a bra by its strap from a curled black finger and calling to me, "Hey, Missus, wotname you callim dis one trouser belonga milk?" Later, for some odd reason, pidgin fell into disrepute. People down south, who had never met an Aborigine, held that it was a racist insult to the Aborigines imposed on them by the whites. Pidgin had evolved long ago purely because an easy common language between whites and blacks, and blacks and blacks of different tribes, was necessary in the name of communication. However, public opinion down south has now been satisfied. Pidgin is still alive and well where it is needed but it has been renamed Kriol (an obvious derivation of Creole) and listed as a native language, so everybody is happy.

The best of the Delissaville dancers, led by Tommy Burinjuk, had previously made headlines when they had been invited to Sydney to dance for a royal visit and Joe believed that the Delissaville singers could have achieved similar fame if they had been given the opportunity.

One day some of these young men invited Joe to accompany them on a dugong hunt. The canoe was chipped and hollowed from a great log and Joe and four others paddled out into the bay, with a sixth man standing poised on the prow with a harpoon spear held ready. They shipped the paddles and floated, completely silent and motionless, for hour after hour. The boys had told Joe that the dugong had marvellous hearing and none would approach if the hunters made the slightest sound.

Eventually one of the unfortunate mammals did venture too near and was instantly harpooned by its adversary diving from the log canoe and plunging the spear into its body. It was towed back to the beach with great hilarity while each man related claims to great past exploits. The Larrakeah, said Joe, could skite as well as any white man. I was invited to the feast on the beach the next night as the spouse of one of the mighty hunters. I thought my dugong steak tasted as much like pork as anything, but I did not really enjoy it. With all the old stories of mermaids, and knowing that these creatures held their young in their flippers and suckled them from breasts in an almost human fashion, I could only think of it with shame as "the night I helped to eat a mermaid".

While Joe and Tas were so busy rebuilding a store, planting a peanut crop, and all the varied duties of the settlement, my life was not so demanding.

Pat looked after the hospital duties and there was an established school and schoolteacher for the children. There were even two lady missionaries to contest the spiritual beliefs of the people. All the other whites lived in new modern houses, recently built, but we didn't even have electricity because it was impossible to wire the humpy we lived in. I couldn't unpack the little gear we did have because there was nowhere to put it. Our food stores and cooking utensils took up two-thirds of the kitchen table and we ate on the cleared space at one end.

If I reckoned that living conditions at Hooker Creek had not exactly been luxurious, I had soon adapted and even enjoyed making the best of what we had. But Delissaville in the middle of the Wet was something again and I had a now-mobile toddler to think of. At night I could protect her from the mosquitoes and sandflies with a net but they, during the daylight hours, were only two of the myriad of flying, stinging, biting insects which bedevilled our lives. I was particularly afraid of the dengue-fever and malaria-carrying mosquitoes and both of us reeked permanently of the oil, dettol and citronella mixture I plastered on to combat them. Grey-green mould grew overnight on almost every surface, particularly leather and wood. Joe's boots resembled grey, fluffy slippers every morning and his plug of tobacco, put into a screw-top glass jar for protection, grew half-inch feelers the first night and filled the whole jar in a week.

I wiped the mouldy covers of my few favourite books every morning with a growing sense of resentment. Small frogs inhabited every one of the many potholes in the cement floor and large green ones flaunted their numbers in the apology for a bathroom.

There was nothing to occupy my time in voluntary settlement work; no improvements I could safely do to our dwelling to pass the time; I had no Nancy for company; and the millions of sandflies made it absolute hell to essay a walk on any of the jungle tracks, not to mention the numerous snakes and pythons. It wasn't that I was a new chum; I'd previously lived in Darwin during a wet season. When Kimmy developed a persistent little cough I grew worried. TB was notoriously endemic among the Aboriginal population in the Territory at that time.

I braved the trip across to Darwin and took the baby to the hospital, where the doctor obligingly gave me a letter to say that she should be moved immediately to a dry climate. I took it straight around to Head Office and waved it in the boss's face while Kimmy, astride my hip, coughed and whimpered pathetically, and I didn't have to pinch her either.

Joe laughed when I got home and told him. Though he enjoyed his job and made the best of the living conditions himself, he resented the imposition of such sub-standard conditions on his family, and had been wondering how he could legitimately ask for a transfer. Anywhere had to be better than this. Two weeks later we were on the road to Beswick in a decrepit old truck of the type the Department seemed to reserve for the inland settlements.

When the jungle by the roadside thinned out and at last the colour of the soil turned from grey to red Joe pulled up, got out, and delightedly kicked up a cloud of red dust. We felt like exiles going home again at last.

> ... When the incident was reported in all the southern papers, the sub-editors changed the appelation of Superintendent to Doctor. After all, who else but a doctor could extract spears right through limbs, sew up deep knife wounds, set broken arms, and treat patients unconscious after a belt on the head with a nulla-nulla?

Nine natives wounded in fight over lubras

DARWIN, Mon.—Nine aborigines were wounded with spears in a fight over two lubras today.

The natives are at the Catfish Settlement, 450 miles south of Darwin.

The fight started when two "offside" aborigines "made eyes" at the lubras who were working in the settlement kitchen.

In a few moments other natives armed themselves with shovel-nosed fighting spears.

Settlement officials tried to stop the fight, but the excited natives took no notice of them.

An aerial ambulance flew two natives to Darwin Hospital—a lubra with a spear wound in an arm, and a man with a wounded leg.

Four aborigines are in Catfish Hospital, where the Acting - Superintendent (Dr. A. Mahood) stitched gaping head, chest, and shoulder wounds.

Three others, believed to have been speared, have gone bush rather than suffer the "indignity" of being kept in hospital.

This report, dated 23 January 1953, was printed
in all the major southern newspapers.

Chapter 6

Beswick:
A Sojourn in Arnhem Land.

Beswick, about sixty miles east of Katherine, on the southern fringe of Arnhem Land, was paradise compared to Delissaville.

Bill and Mrs Grimster made us welcome in the old homestead which had been the original dwelling in the time when the settlement had been a private cattle station, but was now almost past repair. Bill had been one of the original superintendents who had started more than one settlement from scratch, going bush with his family and building living quarters from whatever he could scrounge. They had raised quite a large family. Some were now at boarding school and only the youngest son, Billy, was still at home, reluctantly doing his correspondence lessons while his mother tried to pack their numerous belongings.

Bill had developed TB in the course of his contact with the natives and was being invalided out of the service. I thought it ironic that one who had done so much to establish various settlements, and been obliged always to live in basic makeshift quarters with a growing family, should now be leaving just as good homes with all the latest modern conveniences were being supplied to staff. Mrs Grimster was a capable, motherly, philosophical woman who obviously made the best of everything. We admired them both.

There was the usual problem when folk long-established in a bush home must depart – what to do with their animals? The two dogs were to go to Brisbane with them, but the rest? Joe bought Billy's beloved young stallion and the chooks and I acquired a pet kangaroo, a delightful pup, an enterprising pink and grey galah and a cat shortly to be delivered of six kittens.

Bill showed Joe the settlement duties he must take over until the arrival of a new superintendent, and I accompanied Mrs Grimster to the daily hospital parade. All superintendents' wives performed these duties though none, as far as I know, ever got paid for it; reason given was that we weren't qualified nurses. (When I think back, some of the emergencies we had to deal with would have frightened you-know-what out of any young nurse accustomed only to a modern hospital and a doctor standing by.)

There wasn't any farewell speech and presentation watch for Bill Grimster as he, his wife and son drove away from Beswick in the beat-up old truck we had brought down, overloaded with their personal gear — just a crowd of proper-sorry Aborigines and two young strangers waving until they were out of sight.

As at Delissaville, there were new modern houses for the newly-appointed superintendent and school teacher but ours was the ubiquitous Sydney Williams hut. This one, though, was on a good cement block, with a wide verandah and corrugated-iron walls round kitchen and bathroom and half-height walls round the other rooms. Before Joe obtained louvres to continue the walls up to the roof I could lie in bed in the early morning and enjoy the frequent through-flights of flocks of a hundred or so budgerigars, twittering non-stop and urging me to be up and about.

The yard was fenced off with a neat, white-painted railing fence and Joe had soon planted an extensive vegetable garden and I let my head go with flowers.

The new superintendent, Tom Carroll, and his wife, Des, soon became our friends. Blonde, exuberant Des had a beautiful singing voice and she and Joe harmonised delightfully in the evening singsongs that became our main entertainment. Tom was as quiet and deliberate as Des was volatile and spontaneous and he had developed the perfect ruse to protect himself from the minor, but myriad, work interruptions of the native inhabitants. He pretended to be hard of hearing. He would cup his hand round his ear and say "Eh?" repeatedly, until the supplicant gave up in frustration and sought out Joe instead, so it was Joe who had to down tools and go to the store or wherever.

Joe said you had to hand it to Tom. Some of the store clients or tool-seekers would request one item only, obtain it, and then, just as Joe had re-locked the store or tool shed and gone back to his job, would re-appear with a request for a second, and sometimes even a third, item. The careful check, "Is that all?" never did make much difference to this frustrating habit. Tom just grinned quietly. He held an Army ticket as an electrician and that gave us the impetus to take the great step forward from the inconvenience of pressure lamps and their unreliable mantles which burst with annoying frequency.

A shed already housed a hitherto-unused new engine and big generator; the new superintendent's and teacher's houses and the school were already wired; there was, reasonably close near Maranboy, an extensive dump of abandoned Army gear. Tom and Joe took the truck over and returned with the tray laden down with coils of heavy-duty wire and all sorts of related bits and pieces which might prove useful to an enterprising electrician.

Tom wired our place and the store while Joe took a team bush and cut down trees big enough for power poles and supervised their installation at the required distances. Came the night when Tom threw the switch and Beswick, for the first time, blazed with light, street lights and all, though of course we didn't have any streets as such. Tom, with a lot more experience with bureaucracy than we had had, worked on the philosophy that you didn't tell Head Office any more than they needed to know, which was nothing because, given the chance, they'd arse it up anyway. For a whole month we enjoyed the convenience of electric light and power and then some big-mouthed evening visitor must have given us away.

Out of the blue, three (why three?) Works Department men arrived in two vehicles — two electricians and an inspector. Shock! Horror! It's true then! Next morning they inspected Tom's work and pronounced it perfect and in the afternoon they went duck-shooting. After tea with us all the inspector announced,

"It's got to come down!"

"Why?" we said, "You passed it okay!"

"It's a Works Department job, has to go through the right channels!"

"Oh," said Tom quietly, "well, if we pull it all down in the morning, will you blokes put it up again while you're here?"

They explained it all to us carefully. Money was allocated for certain jobs each year. Then there had to be requisitions and tenders for materials, all of which took time. The cost of their inspection trip, travelling time and allowances cut into Beswick's allocation, which wasn't anywhere near enough for a major job like this one, anyway.

"Sorry, Tom," said the inspector gently, "She'll have to come down. Maybe next year!"

"Okay," said Tom, shrugging resignedly. Then he told them about the swamp absolutely teeming with ducks about forty miles away. If they left early in the morning they could fit in a day's shooting and still be back in Darwin by Friday afternoon. They left early, didn't wait to see us chop down our power poles and roll up the wire.

Tom replied to the terse reprimand from Head Office just as tersely: "Improvement dismantled". He didn't dismantle it, of course, but the paperwork was right.

Our sojourn at Beswick could not be related without mention of "The Master". Beswick school got him first, but he was a talking point on every Territory settlement for the next few years as he passed down the line. He came from Tasmania with his wife and two children and we, in Tom and Des's absence, followed the Territory custom and invited them to tea on the day of their arrival.

The Master's family did not speak, kept their eyes lowered to their plates; occasionally his wife gave a slight nod, a gesture of punctuation to her husband's comments. She was dark, slim, with a sweet face, but what was she *like*? I couldn't get a word out of her, just a nod or a shake of the head in answer to my questions. Her husband addressed himself only to Joe, as if I wasn't actually there. My God, straight out of Dickens, this boyo, and in the unlikely environment of a Northern Territory Native settlement!

He told Joe that they had, on their journey north from Alice Springs, been obliged to sleep by the side of the road, His tone implied that it was Joe's fault that some hostelry had not been sited at the exact place where he desired to stop. He said the climate was not conducive to a white woman exerting herself with housework, so Joe offered to find suitable Aboriginal help on the morrow. "Oh, no!" said The Master, with a shudder, no blacks in his home, certainly not; he'd organise for a couple of Chinese girls to be sent down from Darwin, very suitable in the tropics.

Joe and I dared not meet each other's eyes as we thought of the Chinese families of Darwin, all rich business-people, who were probably next in line to the government in the employment of white staff. To anyone else we could have explained, but it was already evident that The Master would brook no tampering with his preconceived beliefs of Life in an Outpost of the Empire.

His wife, he told us, had a Science degree: I think that was meant to be interpreted as the reason why she could not be expected to mix with us of lower intellect. I caught one fleeting glimpse of shame and despair on her face and my heart went out to her. Neither Des nor I ever did speak to her though; we never saw her once she entered the portals of her house prison.

We rarely saw the children either. We guessed she must be teaching them by correspondence and she'd be busy too with the housework because, when no Chinese girls materialised, the indigenous house-helps refused to stay longer than two or three days.

Joe nicknamed him The Master and it was Joe who had the most dealings with him; Tom resorted to his hard-of-hearing tactics. When Tom was away and Joe in charge. The Master appeared and demanded the store keys. Joe offered to go with him and issue what he wanted. The man fell into a paroxysm of rage and thundered that he was a *teacher,* he was *in charge.* Joe told him to go away or he'd do what he'd never done before to a white man in front of blacks; he'd thump him. The Master jibbered and Joe took a couple of steps towards him and The Master ran like a hare.

We all used to grin when he roared off in his car to Maranboy every Thursday. That was the day he used to phone Head Office to report us. Only Tom was safe.

He reported seeing Joe walking from the store to our house with two tins of Sunshine milk stolen from the store. (Joe had mixed paint in two empty tins.) He reported me for stealing oranges from the store when he saw me give one each to Kimmy and Des's little daughter playing in our yard. (The store never carried fruit; I'd bought them in Katherine.)

He reported Des and me for making two ball gowns on the settlement sewing machine. (The settlement didn't have a machine. Des and I were altering two frocks for the forthcoming Deb Ball in Katherine on her machine on my cool verandah.)

As soon as school was dismissed, that man prowled the perimeters of our fences like a KGB officer on a red hot trail. Naturally, we soon fell into the habit of providing him with clues. The Wednesday that Des and I staggered furtively from the store carrying a huge carton (empty) between us, he couldn't even wait till Thursday, but roared away to Maranboy in such haste that he nearly ran down two little black kids playing in the creek crossing.

Life was never dull while The Master was at Beswick. Finally, a Head Office representative came down from Darwin, had a word with Tom, and The Master was transferred to Yuendemu.

Our stay at Beswick was not without its dramatic moments. The first caught us quite unprepared. It happened in the interim between Bill Grimster's departure and Tom Carroll's arrival. Joe, whom I had never before seen with so much as a headache, developed a high fever with alternate bouts of burning and shivering. At the same time an increasing number of Aborigines reported the same symptoms at my hospital parade or told me of others in their camps too sick to leave their swags.

There was no transceiver at Beswick and the nearest communication with the outside world was the telephone at the Maranboy police station. I sent a boy on a horse with a note to the policeman asking him to contact Katherine hospital to ask for medical assistance.

In the meantime, the bore broke down. How Joe did it I'll never know. His temperature was 104 degrees F when he struggled out of bed and, with the help of two Aborigines, worked a whole long afternoon to re-instate the water supply. He staggered home just on sundown and for the next three days was in and out of delirium.

All I had in the hospital was aspirins. These I gave to Joe and to my increasing number of patients. As most could no longer come to the hospital, I made the mile-long trek round their camps twice a day with a boy carrying two buckets of water and a large tin of aspirins and a pannikin in my hand. Mrs Grimster had told me grimly to avoid sending any patients to Katherine; they invariably died. She said the young doctor there was an unbelievable disgrace and that native patients were just put in a little tin shed out the back and left there untreated. I hardly had time to think of this as I did what I could to nurse and issue stores, while my house girl, Mollie, took charge of Kimmy.

A fortnight after my appeal for help the doctor came. By the time Joe was back on his feet, six of my other patients had died. The doctor didn't seem interested in seeing any of those still sick until I practically forced him to come to the hospital room. Then he began to dispense aspirins, handing each person a pannikin of water, without even rinsing it between patients. I snatched it from his hand and substituted a clean one each time. Mrs Grimster hadn't been exaggerating. I had intended to send back with him a very sick girl who had just had a baby and who I thought might have septicaemia. After I'd seen him at work I thought better of it and dosed her myself from the antibiotic supplies the hospital had sent with him. She recovered. The same afternoon the Carrolls arrived and Tom sent a report of the illnesses to Head Office.

Some weeks later we had a visit from Doctor Cooke, brought out of retirement down south to carry out some tests at the settlement. He had been the Territory's expert on the malaria epidemics which had once been such a health hazard in the Territory. It *was* malaria; I'm glad I hadn't known it at the time, without any quinine tablets in the hospital cupboard and no way of getting any.

One of the advantages of living at Beswick was that our friends, Tom and Moya Ronan, lived at Springvale Station, not far from Katherine, and as Joe had to make a fortnightly trip to town on settlement business, he would leave me at Springvale for the day and pick me up after his work was done.

Tom had won the annual thousand pound prize for the best Australian novel the previous year and was therefore temporarily affluent enough to devote his time to full-time writing. That was how I came to read "Moleskin Midas" as a serial while it was being written. It was also a prize-winner.

Springvale was the first station settled in the Territory, built on the bank of the river of stone from a nearby ridge, with slate floors and long narrow slits of windows through which a gun could be fired against attack. The original settler had also planted Indian rain trees almost a hundred years before, close by a natural spring. In that lovely place Tom and Moya were raising their family with all the untouched bush beyond the door and a library to delight the mind within. Joe and I treasured those stimulating visits and we laid the foundations of a lifetime friendship.

Kimmy had had her first birthday at Delissaville and now, at Beswick, she was an energetic little person always on the go and talking volubly. Often she was the centre of attraction in a group of women and children sitting cross-legged in the shade of a tree in our yard and Mollie used to take her for a long walk every afternoon. I later learned that it was to a billabong when Kimmy handed me a big blue waterlily one day and Mollie told me the piccaninny could swim allasame little fish and had plucked it herself.

When I say Kimmy was already rattling off long sentences, so she was, but in the local Djaun language rather than English. She would address a question or comment to me and when I couldn't understand her, stamp her little foot and repeat it in a shout. I'd have to get Mollie to translate for me, which she did with peals of laughter. Then she and Kimmy would have a little conversation together at which I couldn't even guess. If I'd thought we were likely to stay long at Beswick, I'd have set to and learned the language myself, as I had begun to do with the Walbiri at Hooker Creek, but Joe was not one to work long for a Department he could not respect and where initiative and enterprise were so severely squashed.

He liked the idea of a government job, with its good wages, taxation concessions and holiday fares, but the only government Department with any real credibility to ordinary Territorians was the Animal Industry Branch, head-quartered in Alice Springs and run by the already legendary Colonel Rose. We knew that Colonel Rose always recruited his stock inspectors from young men who were ringers on stations, who knew the grassroots operations of the industry, but there were only about eight positions for the whole Territory and none of the incumbents was at all likely to want to leave. Joe had already considered submitting an application in writing when he had to take some Aborigines to Mataranka one day, called in at the hotel for a beer afterwards and saw Colonel Rose and a companion drive up. Joe had never met him, but now was as good a time as any. He introduced himself and asked to be considered for a job as a stock inspector.

"Where are you working now?" snapped the Colonel with the military precision for which he was famous. When Joe admitted to Native Affairs Department the Colonel threw back his head, roared with laughter, and turned his back on Joe, who retired, somewhat abashed, to finish his drink with an acquaintance. But, on leaving the bar, Colonel Rose stepped aside, looked Joe up and down, tapped him on the chest and said "Write me that letter!" and was gone.

When Joe came home and told me about it, he wasn't very hopeful, but he wrote the letter anyway, setting out his experience as a ringer and his time with the stock and station firm.

A couple of months later Joe was advised that he was to be transferred after his ten weeks' leave, due to begin at the end of November, out of the Top End district to the Centre, to Haast Bluff settlement, and in the meantime, he was to go to adjacent Beswick station, run by the Department as a cattle station, to relieve the stockman there who was due for his holidays. Ted Morey was the manager, one of the Territory's best-known and respected early policemen, and we really enjoyed our few weeks' stay there with Ted and his delightful wife, Kath.

We'd already packed our gear and left it at the settlement, receiving an assurance that it would be waiting for us at Haast Bluff when we got there.

Early storms heralded the beginning of the wet season. Ted left a utility on high ground on the far side of the Waterhouse River which, once the Wet became established, cut off the station for the duration. Thus, anyone who could swim or boat across the river could then drive the rest of the way to Katherine. Two days before we were officially due to leave, Ted got a message on the transceiver that the Waterhouse was coming down in flood.

He appeared at our door just as I was setting lunch on the table.

"Can you be ready to leave in ten minutes?"

We could. We did. When we got to the river the water had beaten us there, despite the reckless speed at which Ted had driven. It was still not more than running-board deep because the river-bed was wide and flat at that location.

Ted was cool and placid as always and I knew that he must have crossed a lot of Wet Season creeks in the course of his long career in the Top End outback. Nevertheless, I clutched my child and closed my eyes as we rolled down the bank and into the water. Joe stood on the back of the vehicle holding our two suitcases clear. Steadily but heart-stoppingly slowly the utility advanced, with a bow-wave spreading behind, until finally, streaming water, it climbed the further bank. In moments we had transferred to the waiting vehicle and Ted had turned his utility around for the return trip.

"Leave that ute at the police station in Katherine," he called to Joe. "I'll fix up for someone to bring it out again."

We nodded and waved and watched until his vehicle had safely negotiated the far bank. The water was appreciably deeper now; we had only just made it. Joe turned to me.

"Good girl. You didn't turn a hair. I reckon some women would have turned on an act and made the job harder!" How could I tell him that I was so paralysed with fright that I couldn't even have raised a squeak!

Joe estimated, as we drove towards Katherine, that we'd made the crossing with no more than ten minutes to spare, and it seemed like a good omen for a beaut holiday. And so it was. We had two years' savings, free air fares to Perth, and were footloose and fancyfree for ten whole weeks.

We dutifully visited all my relations, who immediately noticed a strong resemblance in Kimmy to her maternal ancestors. We partied with friends, bought a second-hand utility and set off for Sydney the week before Christmas.

There was a cool spell when we drove across the Nullarbor Plain but we were highly disgusted at the vandalism and litter we found at every Government tank, placed at strategic intervals to provide water for travellers. Luckily we carried our own water; we'd been too long in the Territory not to do that. There were axe slashes through the bases of tanks, taps wrenched off, and broken bottles everywhere, despite the fact that pits had been provided for litter. We, like all bush travellers in the outback, had always dug a hole to dispose of tins and bottles at every place we stopped for a meal, well aware of the lingering deaths of cattle with the tins they had tried to lick out fixed over their bottom jaws, or the tin over the hoof cutting into the hock and causing a lethal rotting wound. We'd never seen anything like this in our extensive northern travels and it left us with a lasting impression of disgust.

We heard on our vehicle radio that there was a Christmas beer strike in NSW so we bought some cases of South Australian beer and cases of stone fruit from a Murray River fruit depot.

Breaking camp early one morning on the day before our arrival at his family home south of Sydney, Joe and I had a petty argument and I flung out my hand in temper. Unfortunately Joe moved closer to me at the same moment and my hand still held a saucepan. By the time I was introduced to Joe's family for the first time Joe was sporting a prize-ring shiner.

"What happened to your eye?" they exclaimed in horror. "Marie hit me!" said Joe.

Nobody believed him.

They all noticed immediately that Kimmy bore a striking resemblance to her paternal ancestors. We had a wonderful Christmas with Joe's brothers and sister and their families, also assembled for the family reunion, and a day shopping in Sydney to buy the very best available guitar and oil paints and canvas. Joe had always painted in watercolours prior to this but, after our visit to the art gallery, he wanted to try oils. We had a happy half day in Angus and Robertson's bookshop, little Golden Books for Kimmy, art books for Joe, and a selection of favourite classics for me.

We went to a hotel lounge for a drink but they wouldn't let us in because Joe did not have a tie on. He was outraged and still narked when we descended some steps to a rather flash basement restaurant. We had our beers with our lunch but when the bill came they cost four times the price of the same item back in Katherine, with all the overland freight to be considered. Joe said maybe the time had come to add up what we'd spent, as we'd just been writing cheques with gay abandon right across

Australia. We got a bit of a shock when we did. There was just enough left for petrol to get back to Alice Springs and a little over to eat on the way. Sydney suddenly lost its glitter.

When we got back to Joe's folks' place, there was a re-addressed letter for him from Colonel Rose. There was to be a fortnight-long stock-inspectors' school in Alice Springs after the Wet and if Joe was still interested the job was his. I think we sang all the way back through Queensland to the Territory, even though we limped into Alice springs in a ute with no bottom gear and a sapling jammed underneath to take the place of a broken mainspring.

Chapter 7

We Move to the Red Centre.
Joe Becomes a Stock Inspector.

We had a four-week stay at Haast Bluff before Joe's transfer to Animal Industry Branch became effective. The scenery was fantastic but the running of the settlement bore no relationship to what we had known previously. This had been a Mission, only recently taken over by the Department, and the stern-faced missionary and his stern-faced Aboriginal acolytes were still very much in evidence. The place had an atmosphere of "Here we all are, guilty sinners, and if we so much forget ourselves as to grin sideways, a bolt from the blue will probably strike us flat, and serve us right. Amen."

Without ever consciously thinking about it, Joe and I had always belonged to that section of the outback community which rises and begins work at sun-up, or even before if necessary, which believes that anybody who doesn't is either sick or a lazy slob. Because of the fierce mid-day heat in the tropics, it is customary for humans and animals alike to seek the shade for an hour or so and "have a camp" after lunch, the Australian equivalent of the Spanish siesta. The early start compensates for this necessary work break. We hadn't yet run across anyone in the bush who didn't make an early start, but at Haast Bluff the situation was very different.

There, the bell to start work rang at nine o'clock, the time that we usually knocked off for smoko on other settlements. The Aboriginal workers had to go to a church service before starting work, which took anything from twenty to forty minutes. The missionary sent a house-girl for me to employ but after three days I sent her back with a note to say that I didn't need anyone. There wasn't anything for her to do because I'd finished the housework myself long before she made an appearance and I had a sneaky feeling that she was really there to report back if I smiled or laughed out loud.

A second offer of a nursemaid for Kimmy was also rejected; these mission girls didn't inspire me with the confidence I'd had in dear down-to-earth Nancy or laughing Mollie. I didn't care if I wasn't being diplomatic and fitting in; we weren't going to be there for more than a few weeks.

Joe took advantage of all this spare morning time to make sketches and colour-code the changing hues of The Bluff and the ranges in the pastel crayons he had bought in Sydney, a marvellous set of sixty different shades. He made pencil sketches of people and scenery and delightful cartoon characters for Kimmy, who was already reaching for a pencil and trying to emulate him.

Albert Namatjira was living at the settlement at this time and was already quite famous as an artist. What a genuinely likeable gentleman he was! He asked Joe one morning if he knew of anyone going into town, as he had used up his supplies of watercolours and paper and needed more. As our belongings actually had arrived before us, Joe's watercolour blocks and paints were in a box, not unpacked, in our bedroom. He was now quite dedicated to a start in oils so he gave all his watercolour gear to Albert, who was profuse in his thanks. A couple of days later he offered, as Joe wouldn't take any payment, to paint a picture for Joe instead. Joe accepted and asked him if he would paint The Bluff from our backyard. Albert set up his easel to catch the early morning light and in two sittings it was done. Joe insisted that he take twenty pounds for it.

Albert, not to be outdone, noticed two coolamons which I had bought from an elderly Aborigine and asked if he could borrow them. He returned them to me that afternoon with two scenes painted on them carefully signed with his name.

We did have one trip to town before we left permanently. The world premiere of the film *Jedda* was to be shown at the open-air cinema in Alice Springs, and we had met Charles and Elsa Chauvel in the Top End when they were making the film and were interested to see how our acquaintances, drover Wason Byers and policeman Tas Fitzer, would handle their supporting roles. The settlement truck took in a large group of native men because they had been the warriors in the film, and excited expectancy ran high.

That must be the only film premiere in the world where the star of the show had to queue up at the ticket office and pay her way in and then sit unnoticed up the front with her girlfriends and, of course, her co-actors, the warriors. We were absolutely disgusted. All the town dignitaries were being fussed over and nobody gave Rosie Kunoth a second glance. Rosie was part-Aboriginal and had been chosen for the star role by the Chauvels from a group of aspiring local high school girls. She was, after the film had been made, employed as a maid at the Residency. The Resident was the town's first citizen and I would have thought it apt if she had been included in his party but then, on second thoughts, Rosie probably had a better time sitting with her girlfriends up the front. For all I know, she might have chosen to do so. But at the very least, she ought to have been given an invitation ticket to get in!

Back at the settlement next day there was more indignation — from the warriors this time. They had, daily for a full week, run back and forth before the cameras, waving their spears and shouting, and they saw themselves on screen for little more than a minute, barely time to see who was who, and some fellas not there at all, boss. What a letdown!

Colonel Rose had organised for Joe to finish work with Native Affairs on a Friday afternoon and begin with A.I.B on the Monday morning, thereby ensuring continuity of government service.

Housing was at a premium in Alice Springs. We solved the problem by caretaking for families on long-service leave, which suited us because we didn't yet own any furniture apart from Kimmy's cot and chair. Joe signed up for night-school carpentry classes, where he had access to tools and timber supplies, and over time made us all we needed.

The new job suited Joe's talents and aspirations down to the ground. The stock inspectors' school was held every two years, to bring the boys up to date on new legislation, new methods and disease control, and basic veterinary operations, and they worked solidly all day and played pretty solidly half the night too. Joe's new guitar was in demand for his reputation as an entertainer had followed him down from the Top End. These get-togethers for the schools were something like a bush race-meeting in that the stockies only all met together once in two years and they made the most of it.

Each man worked independently in his district, had to make the right decision on the spot and furnish a monthly report to Alice Springs office. There is no doubt that Colonel Rose's leadership was the reason that the branch functioned so well and was held in such high regard in the bush. He demanded and got 100% dedication from his men. He considered the field staff were the important men and the office staff were only there to back them up. His boys would have followed him anywhere and Joe was soon no exception. Niggling and inane government orders from Canberra, via Darwin, meant nothing to "the Boss". He told his boys to ignore them and he'd handle it. "Daily diaries and mileage notations! Bullshit! You've got a job to do, and no time for useless paperwork that contributes nothing to the job in hand. Forget it!". Field officers in other branches meekly filled in data and lamented the freedom of "Rosie's boys".

When the instruction came that all government vehicles were to be returned at the end of a trip to a pool in the town where they were stationed, and a vehicle requisitioned out as required, the Colonel hit the roof. His boys were issued with a four-wheel-drive vehicle, they kept it in tip-top condition, equipment, swag and tuckerbox always on board, ready to leave town at a moment's notice, day or night, as soon as a request for their services came from a drover or pastoralist. As Joe said, when the Boss gave you a job to do and asked "When can you leave?" you didn't answer "Tomorrow morning", or "Four o'clock"; you said, "Give me ten minutes to let my wife know where I'm going", and if you were still around twenty minutes later he bawled you out. The A.I.B boys kept their vehicles at their homes; the pool was for the rest.

From the wives' point of view, we considered ourselves as part of the team too, because Colonel Rose, while running the Branch so efficiently and handling the Canberra paper-pushers so ruthlessly, still had time to consider our welfare behind the scenes. Later, when Joe was stationed at Finke (a day's travel from Alice Springs) and was flown out after an accident, Colonel Rose turned up twenty-four hours later, just checking to see how I was coping and did I need any help. I assured him I was right, but a few days later he sent one of the other stockies down with some town bread: he chopped up the whole woodpile for me and volunteered for any other chores I might have.

After two years with Native Affairs, such efficiency and concern was like a dream to us. Colonel Rose was one of a kind. All stock inspectors did a year in Alice Springs, learning — and learning quickly — how they were expected to perform. Then they were allocated a district. Where there were no vets, they had to become vets; Colonel Rose taught them himself, adding the finer points to the operating techniques most of them had already acquired to some extent in their previous occupations as ringers in the bush. They had to diagnose animal diseases and they had to be right. They had to be able to borrow a horse from a drover and ride into the mob and cut out a suspect pleuro beast, knowing perfectly well that the tough drover, who didn't want to be held up on the road, would pick out the roughest outlaw in the plant for the stockie to ride. Only a capable ex-ringer, well-versed in the wiles of the stock-handling profession, would understand all the nuances of the game.

They had a regular turn at meat inspection at the abattoirs and got their tickets; they became Wildlife Rangers for their districts and were instructed to consider and report on what is now known as Landcare angles — overgrazing, erosion control.

Colonel Rose was a brilliant man who saw both the wide picture and its finest detail, and his boys must see it too or he'd know the reason why. After all, he depended on them to give him the picture; they were a team, and a team with no weak links. Every station in each man's district must be visited, at a minimum, twice a year, whether there was a specific job to do or not. Flying the flag and keeping your eyes open, the Boss called it, and take your wives along too if you like. The government edict that no civilians or family could be carried in government vehicles didn't apply to stockies, the Boss said. If they could give a lift to a station person, or carry some needed gear for them, they were to do it, and don't forget to pick up the station mail and some town bread before you leave and don't arrive there just on meal time. The station people were part of the overall team too and had to be considered in every way possible.

However, we had not broken our ties completely with Native Affairs. We had been in town less than a fortnight when the typist at the Native Affairs office was suddenly called down south to a sick relative. "Can you type?" Arch Richards appealed to me. "Just for a few days, till she gets back." Well, I could type the way most journalists type and Arch was a good scout, seconded, I think, to try and give the Department a better image with the general public, so I agreed to help out.

The girl never did return and I stayed on for nearly a year until Joe was stationed at Finke. I found the perfect day-minder for Kimmy— Mrs McRae— and commenced a relationship that would eventually, over the years, provide all four of our children with a firm but loving grandmother substitute.

I was Arch's secretary, and typist for the other five men in the office. There is no way I would have passed the shorthand or typing tests normally required for government secretaries. Arch soon woke up to the fact that I was drawing squiggles and memorising his dictation. In the end he'd skim the letters I'd already read, comment "Tell 'em such an' such!" and leave the rest to me. I knew the jargon; we saved a lot of time that way.

The other clerks didn't have much to do and they had to write out their notes in longhand for me to copy. I made morning tea for Arch and any of his visitors and horrified the rest of them by putting up a tea roster.

"Our last girl brought us all our tea!" they said.

"And so shall I," I said. "One week in six ... *if* I get mine from you for the other five!" They soon got used to it.

I particularly enjoyed answering the letters from The Master, now stationed at Yuendemu. He reported the superintendent for taking a short-cut through the school grounds to the office and the missionary for stealing beef, and one day Arch got a telegram "Request protection — repeat protection — from Assistant Superintendent".

It appeared that the Assistant had brought the mailbag from town and the wrapper of The Master's daughter's correspondence art lessons was torn. The Master accused the Assistant of tampering with his mail, and that long-suffering individual promptly threatened to thump the bastard. Arch wouldn't let me send the first reply

I submitted, which read, "Your request deemed irregular. Permission granted to Assistant to thump you". He said we could say the same thing in essence but it had to be phrased more diplomatically.

Native Affairs wanted to set up a cattle station on one of the settlements, so a request was submitted to A.I.B from Darwin head office for a report on the suitability of the country and advice on how to proceed. Colonel Rose gave the job to Joe, who had done exactly the same thing voluntarily when he was initially so enthusiastic about the job at Hooker Creek. His plan then hadn't even been acknowledged. I saw the report when the Darwin office sent it down to us with instructions to follow "the expert's" advice to the letter. Nobody in Darwin apparently saw any irony in the fact that they already had a very similar document by the very same "expert" in their Hooker Creek file.

Arch asked me to standardise the spelling and work out the equivalent skin names for all the Territory tribes, which I did with help from Merv, now back at university in Brisbane. Copies were issued from Darwin to all patrol officers and relevant office staff. I got one back myself along with a directive to use this spelling at all times.

That was a good year. We made a lot of good friends and Joe saw a lot of country on his station visits. There was little time to paint, but his sketch pad was always part of his equipment. He was asked to do the illustrations for a booklet which A.I.B issued on poison plants of the Territory, and later for a tourist guide book compiled by a local man.

In January of the following year, a new A.I.B. district was established. Its perimeters were the South Australian border, the Western Australian border, the Queensland border and a line drawn across the Territory about sixty miles south of Alice Springs — home base Finke on the Adelaide-Alice Springs railway line. Joe was to be the district's first stock inspector. Someone gave us the opinion that Finke was the arsehole of the world. It wasn't really — perhaps the belly button. A brand new house was awaiting us. We loaded up our gear, attended some riotous send-off parties, and set forth.

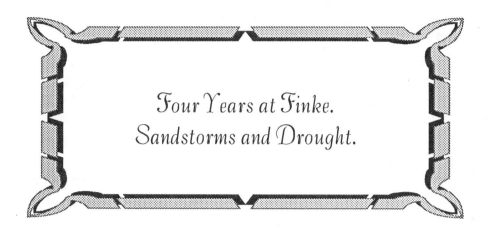

Chapter 8

Four Years at Finke.
Sandstorms and Drought.

The township of Finke was always known as "The Finke" in spite of the fact that it was a mere mile or two from the river of the same name, which was always referred to in full as the Finke River. It flowed, on average, once in five years. The township comprised an old stone pub which had seen better days; a corrugated iron store, ditto; police station; a repeater-station and post office; carrier's yard with caravan; railway workers' quarters; four new railway houses, empty except for the end one which had been leased to A.I.B for us; and the corrugated iron "Hall". The Ghan train from Port Augusta to Alice made the trip weekly, once up, once down.

The township and railway line lay in the hollow of two lines of red sandhills, long since denuded of even the sparse vegetation normal in that region of four-inch average rainfall. The sky was bright blue, the land bright red, and blowing sand a regular hazard. The railway gang's daily job was to keep their section of the line clear of sand.

The Finke was three hundred miles south of Alice, not far from the South Australian border, and we had left town the night before and camped twenty miles down the track to get an early start. It is not wise to travel in the middle of the day in the baking heat of a Central Australian January day, so we expected to find a shady tree and pull up for a few hours for dinner camp. We had to pull up, anyway, when the road disappeared completely under a high red dune, whose outer edge slid down into a rocky creek quite unnegotiable by vehicle.

A couple of hours shifting sand with a long-handled shovel eventually got us through and then we came to a crossing of the Finke River, white sand this time and a good hundred yards or more across. We got bogged in the first ten yards and that meant unloading all our personal gear — two beds, a cot, two armchairs, table, chairs and so on — and carrying it all across to the other side, sinking calf-deep in the burning sand at every step. We sure drained the waterbags with that effort. Then Joe had to go back and dig out the vehicle and nurse it across. It was sundown when we arrived at The Finke.

The townsfolk welcomed us with a meal and the storekeeper lent us a lamp until we had time to unpack our own. We knew we were instantly welcomed into the community when they told us it just so happened that all three of the district associations had just held their Annual General Meetings and had elected Joe as secretary of the Race Club and me as secretary of the Progress Association and the Social Club.

We set aside one room in our new house for Joe's studio, but he didn't get much time to use it with such a vast district to cover. Once the cooler weather came, he had to supervise the frequent trucking of mobs of cattle onto the trains for the Adelaide market at The Finke and also at Abminga to the south and Rumbelara to the north. Rumbelara was only twenty miles up the line but a hundred miles around by road because of the huge impassable sandhills beside the line.

Joe used to drive out of town in his Landrover, wave cheerily to our friend, the policeman, and then double back and climb up onto the single-track line on the south side of the river and drive along on the sleepers to his destination, naturally timing himself so he and the train didn't want to use the track at the same time. I sometimes went with him for an outing. I wasn't really happy when he started going that way on his necessary trips to Alice Springs, jumping off the line at Deep Well. I didn't fancy the long road trip either and preferred to travel up and down on The Ghan but I worried that he might get caught and prosecuted. Joe gave up that route beyond Rumbelara after the night he came round a curve and saw the lights of an unscheduled goods train coming towards him. He had to jump the Landrover off the rails down a steep bank and it buried its nose up to the windscreen in a sandbank. He couldn't very well walk to the nearest rail siding and ask the fettlers for assistance to get going again and he was many miles from any road or track. It took him all night to dig out and sneak back onto the line and home before he was caught in the act.

I have mentioned that the other three railway houses were empty. This was the result of a seemingly irreconcilable impasse between the Railway Department and the Education Department. The Railways wouldn't send families there until there was a school and the Education Department wouldn't provide a school until there were children living there. The Progress Association had been asking off and on for a school for the past fourteen years, to no avail.

In keeping with the environment Finke was camel country. There were hundreds, probably thousands, of wild camels in the district and one enterprising Aboriginal family had acquired an old motor vehicle with a clapped-out engine from the railway ganger, stripped out the engine and hitched their conveyance behind two camels. They carried their water canteens where the engine had been. I could always tell when they were coming in to town because the camels made almost as much noise grunting and roaring as a car engine would have done.

The policeman was greatly envious of Joe's new Landrover. The Police Department had supplied him with a string of camels instead of a vehicle, one or more of which invariably kicked him when he had to change their nosepegs annually. The Finke was a hundred miles east of Kulgera, on the north-south road, and when he was instructed to apprehend car thieves known to be making south the policeman felt himself at a disadvantage both physically and psychologically. The thieves had generally reached Oodnadatta before the camel-mounted cop had even got to Kulgera, and the ribald comments of local inhabitants did nothing for his morale. When the Police Department eventually heeded his plea for a vehicle, the whole town threw a party. This was towards the end of 1956, just before I went to Alice Springs for the birth of our son, Bobby.

Two months earlier I had off-sided for Kitty Colson, who ran the store, during the difficult birth of an Aboriginal child. Kitty had been a nurse and everyone in town depended on her in emergencies.

Lorna, the mother, was in labour all that long night. We had only the light of a kerosene lantern in the kitchen behind the store, and the baby was lying in a transverse position and flipped back every time Kitty managed to turn him. She roused the postmaster to phone for a medical plane at first light but feared that it would be too late. While she went to phone I turned the baby once again and, with sheer luck, timed it with a contraction. Kitty got back just in time to welcome little Roger into the world. The cup of tea that Lorna, Kitty and I partook of ten minutes later was the best we'd ever tasted as the first hint of piccaninny daylight crept over the sandhills. Kitty said firmly, "*You* are going to Alice Springs." I thought I'd better. The three bush births I'd witnessed so far had all included complications.

As it happened, Bobby's birth was even more precipitate than Kimmy's, half an hour from first pain to cup of tea in bed, and the doctor not arriving until after I'd already phoned a message through to The Finke to let Joe know. The postmaster was a bit surprised. He told me they'd had a big party in town the previous Friday when some of Joe's station friends had been in to meet the train and Joe had announced the birth of his son. So I told Bill to tell Joe it was a girl, but to put him right next day.

My return home a week later coincided with a heat spell in central Australia which was the hottest ever recorded. In all, there were fifteen deaths from heat stroke. Birds fell out of the air dead before they hit the ground. Every plant in our garden was charred black.

It was impossible to stay inside the house and we put the baby's bassinet on the verandah and hung wet nappies along its sides, re-wetting them every ten minutes. We filled Kimmy's canvas wading pool in the shade and daily dozens of small birds and their natural enemies, the hawks, packed themselves together against the wet canvas sides, quite oblivious of each other and the humans in and out of the pool.

I'm not sure how we got through that time, but I remember that when Joe kissed me one morning there was a loud crackle of sound and we both leapt apart with an electric shock on our lips. To touch anything metal was really hazardous — refrigerator, cutlery — and at night such contacts were accompanied by pyrotechnical little lightnings. We couldn't even take off a garment without a display of lightnings and a series of minor body shocks. Kitty opened the store after dark. During the day it was too dangerous to risk walking even a few hundred yards in the sun. The railway fettlers stayed in their quarters; the train, if it came, would have to take its chance with the sand.

Temperatures reverted to normal — from life-threatening hot to just stinking hot — a few days before Christmas, and we all celebrated joyously.

We stayed over four years at The Finke and added a second daughter, Tracey, to the family at the end of 1959. Joe, camping alone so often in the bush in the course of his long trips to outlying stations, developed an affinity for the desert country which was never to leave him. It was country you could never take lightly and no land for the unwary traveller, for every year still brought its toll of death to unlucky strangers unversed in the rules of desert travel.

Joe was in the party which found the bodies of two young men, both Scout leaders in NSW, who had bogged their well-equipped vehicle in a sandy creek on the way to Ayers Rock and then tried to walk back to a station in the heat. One of the party let some air out of the tyres and drove the bogged vehicle straight out. Even the smallest local child was taught that you stay with the broken-down vehicle in most cases, but if you must walk as a last resort, you walk only at night.

Among the rewards to those who know the rules is the most glorious winter climate in the world and, to the artist like Joe, the clearest light and the most vibrant colours of any known landscape. The desert thrummed with life and Joe became as adept as any Aborigine at tracking the many small creatures which inhabited it.

An A.I.B house had been built for us in our second year there and when Kim turned five I began teaching her and Lorna's half-caste son, Ian, by correspondence lessons.

When a family with three school-age children arrived to run the pub, I reckoned I could round up the numbers for a school. At that time school was not compulsory for Aboriginal children, but there were a few families I thought I could persuade to camp close enough to town to send their children along. As secretary of the Social Club I wrote a letter to myself as secretary of the Progress Association and offered and accepted the use of the hall as a temporary school. Then I wrote to Darwin with details and listed the names of the school-age children.

I suspected that the education of the Aboriginal children might be considered the problem of Native Affairs Department, so I just used their own whitefella names and happily made up surnames for them. All we needed was a teacher; she could board at the pub.

The new Director of Education in Darwin began to dictate a letter back to me to say that there just wasn't a teacher available to send, but his secretary happened to be a friend from my university days and she suggested that he ask me to take on the job. This took me by surprise, but as it was the only way we were going to get a school, I accepted. Lorna had been working for me while I taught Kimmy and Ian, and she was quite prepared to look after my Bobby and her Roger while I was at school.

The next hitch was that we had no desks. The Director wrote that we would have to postpone school opening until the next financial year. By this time I'd got too enthusiastic for any hold-up so I sent him a telegram to say that we had been donated furniture and to go ahead and send us down the stationery and textbooks. (All except four pupils were on the free-books list.) While Joe made me two large blackboards I borrowed a truck and went to Alice Springs. There, with some discreet help, I stole the desks from the brand-new school — items which I knew from asking a teacher friend, were old desks not in use and designed to be sent back to Adelaide. I also took a crash weekend course from two helpful teachers on how to run a one-teacher school, and they loaded me down with charts and teaching aids. We stole the desks at dead of night and I left town under cover of darkness.

I loved that school. We soon had 22 pupils when the Railways families came. They were migrant families who had to work for the government for two years in return for their transport from the Displaced Persons camps in war-ravaged Europe. From Europe straight to The Finke and with very little knowledge of the English language — they must have wondered what they'd struck!

My pupils were a mixed bag. I had eleven white, eight Aboriginal and three half-caste pupils and Lassie, Ian's dog, who never missed a day in the two years of my career as Head Teacher, Finke State School. Between us we spoke five different first languages, including German, Czech and Pitjanjatjara. Never a dull moment.

We needed a library once the kids could all read. I sent a list down to Angus and Robertson in Sydney and told them to send the bill to the Education Department in Darwin. Luckily the four crates of books had arrived and been covered and numbered before I got the letter from Darwin saying they must be returned— I must requisition through the proper channels. "Too late," I said, "they're numbered and labelled. The shop won't take 'em back!" I don't know how the Director squared it off with the Finance Clerks but, a month later, I also received a box of second-hand children's books from the Darwin office and, not long afterwards, a brand-new kerosene refrigerator to keep the free school milk cold. As Joe said, thinking back to Hooker Creek, sometimes initiative paid off and sometimes it didn't.

The day came when the Health Department caravan and nurses arrived to do the compulsory chest X-rays as part of the TB eradication campaign. The van was parked opposite the pub and everyone in the district was in town, including a ringer from New Crown Station with a reputation for practical jokes. I had my apprehensive kids all lined up when Bill came along.

"Nothing to it, kids," he told them. "The Sister reckons it doesn't hurt a bit! Look, I'll go in first and prove it to you!"

I might have known. That rat went blithely up the steps, disappeared inside, probably pinched the nurse's bottom, and re-appeared, staggering, clutching his stomach, rolling his eyes and moaning lugubriously. In an instant my neat line of kids exploded like rabbits for the sandhills. It took over an hour to round them all up again.

When any visiting clergy came to The Finke I used to ask them to give the children a non-denominational talk, but I drew the line at the missionary from over the border. Everyone knew that this man regularly interfered with the teenage Aboriginal girls in his charge and had fathered a number of half-caste children. The manager's wife of the station adjacent to the Mission, to whom the distraught girls sometimes fled, remarked, "It's a pity the bastard doesn't wear his trousers the way he wears his collar!" He was, Joe claimed, the only man who was able to render me speechless. It was race-time and I was selling raffle tickets to the crowd outside the store in support of the AIM Hostel in Alice Springs when the missionary strode up, almost frothing at the mouth, and loudly denounced me as an advocate of Satan for luring innocent people to commit the heinous sin of gambling. When he was eventually jailed for two years for his extra-curricular activities, no one was surprised.

A piece of dirt of our own was still our ambition. We had saved the required two hundred pounds and regularly entered our names for the Queensland Government Ballot Blocks as the bigger stations were being split up. The lucky winners were financed by long-term, low-interest government loans. We would spend long happy hours poring over the maps of the blocks working out how we would develop them but we were never lucky enough to win one. On two occasions, when local stations were on the market, Joe and Peter Severin, then manager of New Crown, applied for partnership finance from a stock agent firm but were unsuccessful. As a second string to his bow, Joe decided to take the Public Service exam for promotion to a higher grade, by-passed the normal two year part-time study course, studied week-ends and evenings for a month, and topped the Territory with his results.

It only rained twice in our time at The Finke and I missed both showers by being away in Alice Springs. The Finke River ran once after heavy rain further north and the sandstorms occurred with dreary regularity. One blew for three days and, when it was over, the five-wire fence around our yard showed only two wires above the sand. Our woodheap and two forty-four gallon drums of fuel had disappeared entirely. Joe had to prod the sand with a crowbar to find them again. Cleaning the house was a marathon task because the sand penetrated everywhere but inside the fridge — even into closed cupboards and between the bed sheets.

Joe found me standing, wondering just where to begin.

"Pity I've got to take a run down to Abminga, or I could give you a hand," he said.

I just dropped the broom and stared at the wretch. He burst out laughing.

"Honey, you come in every time," he said. "C'mon, this needs organisation. You do the walls and cupboards and I'll do a first run across the floors. Sooner we start, sooner we finish."

It took us two days, carting out buckets of sand, sweeping and re-sweeping, washing floors, walls, crockery and bedclothes, all the time dripping sweat and making heavy demands on the verandah waterbag.

It was a situation being repeated in every other dwelling in town. We all gathered at the pub on the evening of the second day, where the old hands soon launched into their yarns of far worse sandstorms which had buried whole herds of drovers' cattle or half-buried homesteads, until we were forced to admit that we had got off lightly this time.

When Tracey's birth was imminent, the Education Department sent a keen young teacher to take over and there was promise of a new school and teacher's residence for the following year. Despite its summer climatic drawbacks, The Finke was a great place, mainly because of the people who lived in the town and the district. Most of them, unlike us regularly paid public servants, depended on their own initiative to survive and they were resolutely cheerful throughout the long drought, remembering the good times when the sandhills blazed with wildflowers and the country was rich with green grass to the horizon and their cattle brought top prices in the Adelaide market. There was no chance of anyone being forced off their places because they were all now so far in debt to the agents that a forced sale of any of the bare runs would not realise a price anywhere near equivalent to the debt, so the agents waited for the good times too, when their clients would once again become solvent.

Tracey was only three months old when Colonel Rose phoned Joe and asked him if he would like a transfer to Alice Springs, to which he readily agreed. The Education Department wanted me urgently to teach French at the High School and had approached Colonel Rose first. An hour later the Education Department phoned me. Through the window I could see another sandstorm brewing on the horizon and I said yes before the Director's request was hardly out of his mouth.

The chronic shortage of French teachers in South Australia and the Northern Territory (which was then an extension of the South Australian Education Department) was due to the fact that French had only very recently been substituted for Latin as the main language to be taught in schools, so there was a surfeit of Latin teachers but a scarcity of qualified French teachers. A foreign language was a prerequisite for University entrance to all faculties, so finding a teacher for the exam class was important. I was in the right place at the right time.

There was a slight hitch about an Alice Springs house for us. When Colonel Rose approached the Housing Officer for a house for us he was told there was none available. By coincidence the Housing Officer was the very same man who had been held responsible for the Hooker Creek incompetencies. There was some collusion between Colonel Rose and the Director of Education and an appeal was sent to the Minister of Territories in Canberra. The Housing Officer's secretary told us later that Paul Hasluck sent a terse telegram "Find the Mahoods a house!" Suddenly we had a choice of four empty houses and we chose one on a large block across the Todd River facing a small park with an unrestricted view of the Mt. Gillen Range. Mrs McRae welcomed Bobby and Tracey with open arms.

Chapter 9

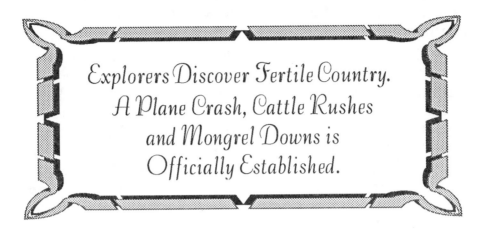

Explorers Discover Fertile Country.
A Plane Crash, Cattle Rushes
and Mongrel Downs is
Officially Established.

Alice Springs had at last submitted sulkily to the fact that it was going to become a tourist centre whether it liked it or not. The town was divided into those who continued to ignore visitors altogether or eyed them suspiciously and those with the enterprise to see that tourists represented potential income. The two hotels were modernised and extended. The tariff was raised accordingly so that the previous station clientele could no longer afford to stay there on their trips to town, and found it more economical to buy their own houses in town.

There were two art galleries and Aboriginal paintings and crafts were at a premium. The local amateur painters formed a society, of which Joe was one of the foundation members, and part of the wide verandah of our house was screened off for a studio. Although Joe was out of town on his job for three-quarters of his time, he was able to paint regularly and won a number of competitive prizes in Alice Springs and Darwin. Requests to paint portraits, design posters and Christmas cards, and paint landscapes grew to such an extent that he began to consider, if a property of our own remained beyond reach, an alternative career as a professional artist. He did a cartoon painting of Ayers Rock with two Aborigines in the foreground, one saying to the other, "Bill Harney reckons it's got some tribal significance!" for Peter Severin at Curtin Springs, and Peter hung it in the store he had now opened to cater for tourists on the way to The Rock. (Twenty years later we were to see it copied rather amateurishly and sold to *The Bulletin* magazine as a cartoon.)

Under a government scheme we were able to buy our house very cheaply so that if Joe did leave the Public Service we would have a base for the family. After two exhilarating years at the High School, I was again pregnant and resigned for a spell of just being home with the family. It didn't quite work out that way because I became, after another emergency, the temporary School of the Air teacher for the station children on correspondence, as well as taking a daily lesson after school for the Junior French classes from both the State and the Convent schools at the request of their parents.

Jimmy was born in June, and I came home from hospital to a scene you didn't often see in a town backyard. I'd asked Joe to go along to the Government Disposal Sale to buy a lawn-mower for our newly planted, extensive and now exuberant back lawn. He came home with forty-eight wooden school desks, circa 1930, which, for want of anywhere else to put them, were stacked all over the lawn. Joe said the timber was a gift and he didn't bother with the mower because obviously he wasn't going to be able to get at the lawn for some appreciable time. I could never find an answer to Joe's logic. He did accede to the requests of some of our station visitors for desks for their children who still had to do their schoolwork on the end of the kitchen table, and I claimed half a dozen myself. However the heavy timber of most of them did eventually make the workshop benches he'd envisaged for our station property.

In March of that year, 1962, Joe, as District Stock Inspector, accompanied an expedition to try to find a suitable stock route to bring cattle through from the Kimberleys of Western Australia to the railhead at Alice Springs.

The pleuro- and tick-free stations of the southern part of the Kimberleys were grossly overstocked, unable to market their cattle to Wyndham or Derby, which both had a hinterland of infected country. They were also unable to send them on the six months long walk south via the Canning Stock Route, where many of the spaced wells had caved in because the timbering had been removed by nomadic bands of Aborigines and water supplies could not be relied upon. Some new outlet through pleuro- and tick-free country was urgent. Bill Wilson, brother-in-law of Margaret Doman, owner of Billiluna Station, approached AIB for permission to try and find a track through the Tanami Desert region to meet up with the old now-abandoned Tanami-Granites goldfields track.

Two Centralian pastoralists, Milton Willick and Bill Waudby, would buy breeder cows if they could be brought through on surface waters after the preceding excellent wet season and both planned to join the expedition. The four men, together with Jim Richardson, the local manager of Elders, and Bob Sprod, owner of the Adelaide business which supplied Bill's Toyota Landcruiser vehicle for the trip, followed the track as far as The Granites and then headed due west on a compass bearing into the desert, aiming roughly for Balgo Mission south of Billiluna.

They carried a portable transceiver and at each night's camp they made contact with Milton's wife, Phyl. I managed to tune in to the same wavelength on an ordinary radio. Joe later wrote the story of the expedition, "Desert Venture", for a government-produced magazine on Australia's Territories. Contrary to what might be expected in the desert, they struck heavy rain and showers for the first three days and all three vehicles were frequently bogged.

The aeronautical map showed only one hill anywhere near their route, marked "Position Doubtful" and made no mention of the rugged sandstone hill only 30 miles from The Granites. They covered 45 miles the first day and 56 miles the next day, but camped for the following two days in the hope that the country would dry out a bit. When Joe's Landrover would not start they traced the trouble to the rotor arm and spent five hours carving out a mulga wood substitute with hacksaw and penknives. They eventually arrived at Balgo after a week's travel but although they had found various waters none were sufficient to water a large mob of cattle.

Bill Wilson chartered a small plane from Alice Springs and he, Milton, Bill Waudby and Joe flew back along their track mapping the surface waters they could see to the north and south of the wheel-tracks. Some lakes looked blue and salty, others red and fresh. The next job was to find them on the ground. Three expedition members flew back to Alice in the plane and Bill Wilson, Milton and Joe set off on the return trip with the three vehicles. The swamp on their original track 35 miles out from Balgo would do for the first watering point. Thirteen miles further on they left their old track and made well south towards the Lewis Ranges, where they had seen good waters in the valleys from their vantage point in the plane. Here they found about three miles of small waterholes. Bill decided to take a chance and radioed back to Billiluna to get a mob of a thousand cows ready to leave. Here they followed a ritual common to many explorers; they named Lorna Springs for Bill's wife, Mt. Phyllis for Milton's and Mt Marie for me.

It wasn't all beer and skittles. One full day was spent following land contours and bird flights to finally come upon a water which, in Bill's words, "If you camped on it for three days with a galah and a mouse, you'd have to give them both a dry day!" But they found and named Bullock's Head Lake and, with great difficulty, Lake Ruth, and by then they knew the trip was possible. Bill radioed Billiluna to start the cattle off. He had already set two well-sinkers to work at a half-way point between Chiller Well and The Granites on the established Tanami Track and they had struck water at 40 feet.

The rest is history. After a day's spell in Alice, Bill went back and dragged a heavy chain along the route between waters to give the cattle a track to follow and Milton and Bill Waudby transported their horse plants to The Granites to meet the mob there. With some variations this was to become the stock route linking the Kimberleys to the Alice Springs railhead. The journey was also to mean the eventual realisation of our long-held dream for a stake in the land.

On the return trip the three men had discovered that to the south of the spinifex and low scrub country of their original track lay unexpected wide plains of Mitchell and Flinders grasses, belts of mulga timber and fresh water lakes. It was, as they were to discover later, described in the 1900 diary of a geologist-explorer named Davidson as "country of such splendid aspect that it is hard to believe that one is truly in Central Australia".

Round the campfire on their last evening together, they decided to apply to the Lands Branch to take up the area as a cattle station in a four-way partnership with Bill Waudby. They were all experienced cattlemen and knew the potential of good country when they saw it.

When Joe described what he had seen, I thought back to a story Wason Byers had told us when we lived at Beswick. As a young man he had thought it provident to avoid a police confrontation by disappearing for some time and he had taken horses, packhorses and stores and headed south from Gordon Downs into the desert. He told us that he had spent two years there before returning to civilisation and had travelled through wonderful cattle country of ironstone ridges and grassy plains and had met groups of exceptionally tall Aborigines, seven feet tall and their women six feet six or eight inches, who, when he tried to ascertain their country, had cracked their knuckles and pointed with their lips to the west. He said that nobody would ever take up the country though, because it was just too far from civilisation to be practical. I had twice before heard of this tall tribe, once from Uncle Jack, who had encountered three men from south of Balgo and measured one of them as seven feet four inches; and again from Jimmy Hazlitt, who had met a family group down near Lake White and the lake which bears his own name today.

It wasn't easy to obtain the requested lease and the considered local opinion was "You gotta be crazy - that mongrel country out in the Tanami!". The Director of Lands Branch refused the application in early 1963. Their inspection consisted of a charter flight of approximately half an hour over a small area and no ground inspection at all and the report was a mere half page to the effect that the land was not suitable for cattle grazing.

Although Colonel Rose was no longer with A.I.B he was more than interested in our plans and quite contemptuous of the Lands Branch assessment. Paul Hasluck was just about to retire from his position as Minister for Territories, so Colonel Rose acted fast. He told the Minister the facts, snorted that the four practical cattlemen who wanted the lease would know good cattle country a damn sight better than the Lands Branch silvertails who hadn't even seen it, and the upshot was that the Minister over-ruled the Lands Branch edict and 1620 square miles were granted to us as a Ministerial Lease in late 1963.

We named it Mongrel Downs, not after the Administrator or the Lands Branch Director or the charter pilot who delivered two watermelons instead of the bottle of whisky requested, as various rumours claimed, but because so many bushmen who claimed to know the outback well insisted on referring to it as "that mongrel country out in the Tanami".

The arrangement was that Elders would lend the money to develop the station and Bill Wilson would go guarantor. (Bill was an established Elders client and he and his wife, Lorna, already owned a number of other stations, in Western and South Australia.) Joe was to be the manager on the block and Milton and Bill Waudby would contribute gear and labour. However, as Milton and Bill both had their own places to run, this was not really workable and after the first year they were both bought out for a nominal sum, leaving Joe and Bill Wilson equal partners in the venture.

When the proposal looked like succeeding, Joe did not agree to it until he had discussed it with me. He put it to me that we both had Public Service jobs with a secure future and regular holidays; we had a family to educate; everything on our horizon was rosy as it was. On the other hand I must understand the isolation and realise that Joe would not often be at the homestead with me; I would have to teach the children myself; nothing was established there like all the other station homesteads I'd known. He would accept the proposition only if I wanted it whole-heartedly too. Now it was no longer a dream where we considered only the bright aspects; we must be practical and think of the broader picture and all the risks we would be taking.

He sat at the kitchen table with a serious face, thinking he was giving me no clue as to how he had weighed up the options himself; it all depended on me. I started to laugh. One of Grandma's sayings came to mind, "Don't cross your bridges before you come to them!" so I didn't even bother to think of how I'd handle the isolation. I said, "Honey if we opt for solid old security and retire on a pension at sixty-five, I'll bet we'll look back and regret that we didn't have the guts to accept a challenge. Whatever happens, at least we'll have had a go".

He jumped up so quickly he knocked his teacup over, and grabbed me and whirled me round and round the kitchen. I was a bit puzzled why he'd even had to ask; after all, this was the chance we'd talked about and dreamed of for years. Did he doubt me? I'd never doubted him for an instant. I brooded about that for a while and finally told him that I was disappointed that he hadn't enough faith in me to know perfectly well how I'd react when the time came. He had known; he hadn't even thought I mightn't be as enthusiastic as he was until some of his workmates began to suggest that it might be a bit rough on a woman and kids, that their own wives wouldn't have a bar of it, that he ought at least to give me the chance to decide, and these comments had influenced his actions. I privately decided that his acquaintances were probably irked that Joe had an opportunity they'd have liked themselves and one man in particular, well, I could have told Joe a few things about how much consideration he had for his wife. If he had an executive job at the Arctic Circle he'd have kept her in an igloo and if she complained of the lack of company he'd probably shove a polar bear in with her.

As Nancy Barrett, the long-term and much-loved School of Air teacher, had returned at the beginning of the 1963 school year I found myself back in my old job at the high school. When the lease was finally granted, Joe planned to resign at the end of the year but common sense ordained that, until a permanent homestead site could be found and some amenities established, his family should remain in Alice Springs.

Bill Wilson had not been idle. He had employed contractors to put down and equip two bores between Balgo and the as yet unsurveyed Western Australian-Northern Territory border, which he sold to Balgo for the nominal sum of one shilling each. He had planned three more bores on the stock route through Mongrel Downs and two more between our boundary and Chiller Well. Because Billiluna was so heavily over-stocked, he had drovers bringing mobs through on a combination of surface waters and new wells, and Joe, as the District Stock-Inspector, had to oversee their progress on the Territory side of the border.

However, Bill's next problem became centred, not on the welfare of the travelling mobs, but on incidents taking place at Billiluna itself, where the cattle had to be mustered. The ladylike and well-educated wife of the manager had acted quite out of character when she had discovered her husband's dalliance with a part-Aboriginal girl. On his return from the stock camp to the homestead she had greeted him with a barrage of .22 bullets across the doorstep as he had entered and followed it up by stitching a line of bullet holes across one Toyota door as he leapt for the other. Then she went back inside and sent Bill a telegram advising him that the whole family would be leaving Billiluna just as soon as transport could be arranged.

Bill called one evening, showed us the telegram and told us the story. "I'm stuck," he said, "there's no one available who knows the place and can do the job. It's only three months before you're due to leave A.I.B, Joe. Will you take it on if I can square your boss? If we can get one last mob away this year, I'll find a new manager by Christmas." Joe agreed.

One drama seemed to follow another. Bill chartered a plane for a return flight for Joe and himself from Alice Springs to Billiluna and Bill Waudby came out with the pilot to discuss buying the last mob of cattle. On the flight back to town Bill suggested a deviation towards Lake Ruth where a contractor known as Dangerous Dan was erecting a windmill and stock-tank on the recently-drilled bore. Bill hoped to see from the air whether it had been completed or not.

The dry bed of Lake Ruth was in sight when the pilot suddenly yelled in consternation,

"We're out of fuel!"

"Make for the lake!" shouted Bill.

But the glide path was too short and they hit the ground a hundred yards short of the smooth lake bed. The plane bounced on its nose and ploughed along a sand ridge before flipping over on its back. Bill Wilson struck his head on the portable transceiver he was carrying on his lap and was knocked unconscious. They were all suspended upside-down in their seat-belts. With a superhuman effort Bill Waudby wrenched off the buckled door of the plane and they crawled out, pulling Bill Wilson after them. "She won't catch fire," muttered the pilot, "there's no fuel!" (What had happened, it was later established, was that the fuel tank cap had come off and the fuel syphoned out in flight.)

They revived Bill in the shade of the ti-tree bushes by the lake-side and then retrieved the portable transceiver and the battery from the plane and were able to contact the Flying Doctor Base in Alice Springs with the news of their predicament. By this time it was mid-afternoon and they were all in shock and desperately thirsty. They lay in the shade, miserably aware that they were a normal two days' travel by road from Alice Springs, but at least Bill's deviation meant that they were near a road of sorts and not crash-landed in the featureless desert to the south, which would otherwise have been the case.

By sundown their thirst was so acute that they decided to walk the approximate four miles to the bore in the hope of finding the mill completed and pumping. It was moonlight and there was a track to follow most of the way. They did not speak, with only one thought in mind as they plodded on, and Joe said they were soon strung out like perishing bullocks going in to water. When the silhouette of the mill finally came into view the pace quickened and Joe reached the tank first. He climbed up the side and looked over. His heart fell. In the moonlight he could clearly see the bolted plates of the base of the tank. He leant closer and his face was suddenly immersed in water! He said afterwards that he had dropped the level of the water in the tank by two bolts before the others had even climbed the side.

With renewed heart they hunted around for the contractor's rubbish dump and dug until they had unearthed some old fruit tins and scraps of wire to make billycans to carry water back to the crash site. Once there they had no trouble falling soundly asleep in the sand beside their campfire.

I got word of the crash while I was still at school. The only person in town who had ever been in that area was Robby Robinson, whose partner had drilled the bore. By nightfall Robby had organised a rescue vehicle while I made sandwiches and thermoses of tea so he and his co-driver wouldn't lose time preparing meals on the way. Robby estimated he could reach Lake Ruth in 24 hours, with his mate Tas to co-drive.

The next morning a search plane located the crash but did not land as they expected on the smooth lake bed (as our chartered planes did until we made an airstrip). Instead it dropped a storpedo and flew away again. They had to walk a mile to pick up the storpedo, which contained no water, but salt tablets instead — just the thing to increase their thirst. They opened some tinned food and ate, though none of them were yet feeling hunger pangs.

In the late afternoon they walked again to the bore to replenish their water supplies and got back to their camp wishing there had been a packet of tea in the dropped supplies. They banked up their fire and settled down for their second night at the crash site, only to be awakened about midnight by Robby's arrival. He had an ice-box of cold beer but all they wanted was tea. The rescuers wouldn't wait till morning to leave and they got back to town with the rescuees in the same time span that it took to get out to them. While infinitely grateful to Robby and Tas for their efforts, when Joe showed me his bruises he said he got the wide one round his waist from the seat-belt in the plane and the rest from the tossing around on the ride home.

The next few days were hectic. Bill chartered another plane and he and Joe flew to Billiluna again where the situation was even worse than they had feared. True to her word, the manager's wife had organised her family's sudden removal, leaving the cook alone and in charge at the homestead. Given such a golden opportunity he had invited three no-hoper mates to bring down a cargo of grog from Halls Creek and it was on for young and old.

Bill and Joe arrived to find the four whites paralytic drunk and a good percentage of the camp Aborigines also drunk and sick. The store had been ransacked and the methylated spirits supplies for pressure lamps and medicinal purposes had been taken to drink when the beer and rum ran out. The whole place was neglected and filthy. It had been a wild old party for days.

"We manhandled the four white blokes onto the back of the ute," Joe told me later, "and tied them down and drove them straight to the Police Station at Halls Creek. They shouted and cursed the whole way. One fellow took a particular set on me; said if it took him the rest of his life he'd track me down and put a bullet in me for spoiling the spree." At Halls Creek the men were immediately charged and jailed for breaking into the store and for supplying the Aborigines with liquor, then a criminal offence.

Bill and Joe returned to the station, Bill to track down the stock-camp and take them overdue food supplies, which under normal circumstances the manager would have done, and Joe to try to restore some order at the station. The Aborigines were shame-faced after their outbreak and only too willing to co-operate.

In the three months that he managed Billiluna Joe supervised the delivery of the last mob to the drover, did a stock-take on the workshop and store, repaired neglected gear and made out the usual pre-wet order for a 12-month supply of estimated rations, clothing, blankets, saddlery, tools, etc. required to run a station efficiently. He said he suspected previous managers might just have automatically re-ordered year after year the previous year's order because he found enough blackstrap molasses to pave an airstrip stacked away at the back of the store, along with carbide supplies long out of use since pressure lamps were invented.

In the meantime Bill had teed up a new manager, Gerry Adamson, to take over by Christmas so Joe could come back to Alice Springs and get organised to do the necessary preparatory work of finding a homestead site and setting up watering points and yards in virgin country before the initial cattle herd was introduced to Mongrel Downs.

It didn't quite work out that way. The last Billiluna mob had set out very late in the season and the weather was already untenably hot and dry and the cattle toey from the start. Passing Balgo Mission the perishing herd got a whiff of water and rushed, quite beyond the droving team's ability to head them off. A thousand head of thirsty cattle descended on the Mission compound, rudely interrupting the slow tenor of a drowsy summer day.

Five head crammed themselves inside a corrugated-iron bathroom until it collapsed about them; two more claimed the church font; three pushed into the office where Father Mack had been catching up on paperwork; a woman carrying a billycan of water tossed it towards the dozen head thundering towards her and ran screaming for cover. It was a hectic afternoon. At Billiluna Joe got the message of the rush and hastily set off with the Billiluna ringers to help get the mob together and on the road again. He passed them on his way in to Alice Springs a couple of days before Christmas, across the border now on Mongrel Downs country but still travelling reluctantly.

We had a traditional family Christmas Day. On Boxing Day Bill phoned. Problems. The cattle in the drover's mob were continually trying to break back, waters ahead had dried up faster than expected, the weather was now so hot that the exhausted men and stock could not possibly cover the daily long stages between drinks. To push on was to invite certain catastrophe. The drover was holding the cattle on good water but was not prepared to take them any further.

That was how Mongrel Downs came to buy its initial herd from Billiluna a good six months ahead of plan. Bill said he would send the Billiluna stock-camp over. All they had to do was stop the now-depleted but very determined mob from walking back to its familiar haunts on Billiluna, where the wet season storms had already begun and the hint of rain on the wind taunted the homesick herd.

Joe left Alice Springs with a loaded Toyota — bags of flour, cartons of dry rations, water and fuel drums, swags and saddles and a few sheets of corrugated iron. "Well, that's the store and the homestead," he said with a grin as he tied the load down. The three Aborigines he had brought to town with him for a spell climbed on top of the load.

Bobby had just turned seven and had persuaded me to let him go with Dad to help him start the new station. He stood with his miniature saddle on one shoulder and his pup, Patch, clutched under one arm, almost too excited to say goodbye. Joe said he'd find some way of getting him back for school next year.

We waved them off in the late afternoon. On the first of January, 1964, Joe arrived at the camp and took the cattle in charge. Mongrel Downs was officially established.

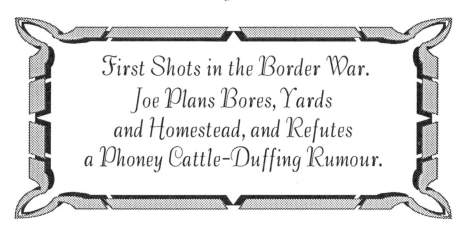

Chapter 10

First Shots in the Border War.
Joe Plans Bores, Yards
and Homestead, and Refutes
a Phoney Cattle-Duffing Rumour.

The cattle had split up into small mobs, all obsessed with the idea of following the known waters back to the west. Joe and Bobby found the Billiluna camp, with head-stockman Les Roche, near Bullock's Head Lake. For the next few weeks they rode continually, turning the cattle back. Some, well on their way home, were picked up and held for a time in a small gorge in Mt Phyllis, roughly fenced across the entrance to hold them while others were added to their numbers.

Robby's partner, Ferdy Bergman, was drilling for water in a belt of bloodwood timber close to the approximate State border. The periodic violent wet-season storms were laying surface water sufficient for the cattle for a time but Bill hoped to have a string of bores operating along the stock route by the time the next dry season mobs from Billiluna began to walk through. Geologists from the Territory Water Resources Branch chose three sites in each desired area and the Government paid the driller if they were duds and the station paid for the successful holes. Joe had a lot of faith in Ferdy's expertise at choosing sites and offered him a double or quits price for tries in key areas where the geologist's sites had proved duds. (Over the next two years three of Ferdy's sites proved successful).

Gradually most of the cattle were pushed back eastwards towards McFarlane's bore, Wild Potato bore and, fifteen miles further on, the bore near Lake Ruth, soon to become Homestead bore.

In mid-February Gerry Adamson passed through on his way to Alice Springs and picked up Bobby for school. The Billiluna stock-camp continued to track-ride the western border country and Long Johnny became Joe's offsider on jobs on the eastern end of the run. Long Johnny was a bit of a loner. As a child his extremely tall family had walked into Balgo from somewhere down the Canning and they did not belong to either the Balgo or Billiluna tribes. Johnny was a smart man and could be relied upon to work by himself. He and Joe had an agreement that neither would go looking for the other until they were two days overdue back at their camp. When I demurred at that, Joe explained,

"Look, Hon, it doesn't make sense to give up following tracks just to get back to camp by sundown if it means losing the cattle. There's all sorts of unexpected reasons can make a job take longer than you first estimate. The bore's not pumping when you get there, you've got to fix it; the vehicle breaks down, takes time to fix. Don't worry. If we got in real trouble and needed help we'd light a smoke for each other."

I knew that, in that desert country, a column of smoke could be seen from many miles away, and I had to be content with that.

The first incident of "The Border War", which was to enliven much of our time at Mongrel Downs, concerned Bloodwood Bore, which had been drilled on a site we believed to be just inside our western boundary. The West Australian border had been surveyed many years before only part way down into the Kimberleys from the north. The southern part was surveyed only from The Bight to the parallel of latitude marking the border between South Australia and the Northern Territory and there was a gap of six or seven hundred miles yet to be surveyed before the West Australian border was complete.

Modern astrofixes had shown that the early surveyors were sometimes as much as three or four miles out from the true meridian. Joe and Bill Wilson took it for granted that the original surveyed lines would eventually be continued until they joined up, in which case Bloodwood Bore was at least a couple of miles into Mongrel Downs country. Father Mack opted for a true astrofix survey when it was eventually done, which would put a bend in the border but almost certainly give him a free bore drilled at our expense.

Joe never did really know what had caused the bad blood between Balgo and Billiluna in the first place and, unaware of any brewing trouble, he had Dangerous Dan, the contractor who had erected the mill and tank at Lake Ruth, on a contract to do the same at Bloodwood. A watering point was urgently needed for our cattle being brought back from their abortive attempts to return to Billiluna.

On receipt of a cryptic telegram from Bill Wilson, Joe drove to Bloodwood, where Dan had just finished erecting the windmill. Shortly afterwards, Father Mack arrived with the Mission grader-driver and the new young policeman from Halls Creek, spick and span in uniform and wearing a gun on his hip. They parked in the shade and came striding over to where Joe waited in Dan's camp. "This bore is on Mission land, Joe," announced Father Mack. "The mill will have to come down!"

Joe thought quickly of the two other equipped bores on Balgo, paid for by Billiluna and sold to Balgo for a shilling each in exchange for right of way through Balgo for Billiluna cattle going to market. "Okay," he said quietly, then raised his voice. "Right, Dan, pull her down!"

This was a reaction the Balgo group obviously hadn't expected. They retreated to the shade near their vehicle and watched the first sail come fluttering down. A quarter of an hour passed and all the sails were down. The grader-driver walked over slowly.

"Father says it doesn't *have* to come down, Joe. We could find a compromise. Will you come over and talk about it?"

"He wants to talk to me," said Joe, "he comes over to me!"

The grader-driver retreated. The head was nearly dismantled before the three men walked over.

"Look, Joe, it needn't come down. I'm sure we can work something out," began Father.

"No, we can't," replied Joe. "*You* said it had to come down, so down it comes! Keep a-goin', Dan!"

The visitors drove away. Joe could never work out why the policeman was there, gun and all; he hadn't said a word.

Dan pulled down the mill and loaded it on his truck. Seven miles further east the driller put down another bore near Lake Alec and Dan erected the mill there with the tank and troughing. From my experience I could have told Father Mack that Joe *always* called a bluff. If he had hoped to purchase an equipped bore at Bloodwood for another token shilling, he sure went the wrong way about it.

Father Mack's next move was to advise that entry on to the Aboriginal Reserve needed a permit from the Perth Aboriginal Affairs Department, which Joe had a feeling he wasn't likely to get, seeing it would be dependent on Father Mack's recommendation. Legally, he could no longer follow our straying stock into Balgo country. Balgo bred and sold horses but did not run cattle and had always bought their meat on the hoof from Billiluna. It was nearly a hundred miles from the border to the Mission on a road which had been made by Bill Wilson's enterprise. It was also the only road between Billiluna and Mongrel Downs, and the only way the Kimberley cattle could come. Joe had already told Father Mack that the good grass plains country continued across the border into Balgo and would support cattle but, until Balgo got its own herd, Joe didn't intend, legal or not, to contribute to the Balgo butcher-shop. A new element was added to the track-riding, that of wiping out horse tracks in too-obvious places.

A compromise on road access was reached later when Father Mack realised that, by travelling on the same road through Mongrel Downs country, the Mission truck could pick up supplies from the Alice Springs railhead, a cheaper and quicker way than the previous boat cargo to Derby or Wyndham and then road transport via Halls Creek. War or not, there are certain rules to be followed in the outback. The Mission personnel always called at the homestead for a cup of tea or occasional help with a vehicle breakdown and carried mail for us, and we did the same when passing through the Mission country.

In May I had a week's school holiday and my friend, Gwen Barnes, agreed to mind the two younger children while Kim, Bobby and I went bush. Joe came to town to pick us up. He was deeply tanned and bearded for the first time in his life. It was an exuberantly brindled beard with strong overtones of red, in stark contrast to his long dark hair, and he looked like a bushranger. I persuaded him to keep it for another few hours on the grounds that love-making with a bushranger would be a romantic new experience for me.

I could hardly wait to see the country and the two-day journey seemed interminable. Joe pointed out landmarks to us and it was easy to see that he was completely at home in that land of red sand, grey mulga scrub and huge anthills, where the horizons seemed to stretch to infinity. To me, its very vastness was awe-inspiring and still a little frightening; we had not seen a vehicle or a homestead roof since we turned off the road sixteen miles north of Alice Springs the previous day.

We arrived at the camp in the late afternoon and found it a hive of activity. Joe had built a big bough shed and a temporary yard near the Lake Ruth bore and Dan's caravan was parked nearby. His wife, Bev, was supervising shower baths for their three children under the overflow pipe of the stock-tank, strategically situated on the other side of the tank from the bough shed. Ferdy and his offsider were there, waiting to check with Joe on a new bore site. Long Johnny was hobbling our horses. Bill Waudby had stopped by on his way to Billiluna with drover Bob Savage, who was to take delivery of the next mob of road cattle.

Bev had a great boiler of stew bubbling on the camp-fire and when all had had a clean-up at the overflow bathroom we sat down in the growing dusk on our rolled-up swags and consumed great platefuls of tasty stew with fresh town bread in place of the usual damper.

The bore was in a belt of mulga and desert ghost gums, an attractive place for the homestead. Joe drew me a mud map in the dirt showing where he would site house, workshop and yards and fence off a mile-square homestead enclosure. Kim and Bobby spent most of their time with Long Johnny and the horses. We drove across country one day to where Ferdy was drilling at the new site at Pat's Swamp and took a run, again across country, to MacFarlane Hills, still marked as Position Doubtful on the map. It was good country and a bore had already been equipped there.

The four days passed too quickly but I had to be back for school, so Friday morning found us rolling our swags for a lift back to town with Dan and his family, going in for stores and a break. Joe squashed my argument that I come out and camp in the bough-shed until the house was built because he said he would seldom be there and it would mean an extra worry for him to have to think about us as well as the cattle. Being unable to contribute to any fencing, yard-building or track-riding, and mindful of the numerous snake tracks near the bough shed, I didn't push my case. I knew it was only a matter of a few months until I could move out permanently because Joe had already arranged for his younger brother, Ken, who was a Sydney builder, to put up the house for us once the materials were on site.

I was fairly single-minded on the issue of a decent house in which to bring up kids in their formative years and Joe agreed with me. This had been one of our stipulations when we joined the partnership. We had seen too many bush families raised in dirt-floored tin sheds with the result that the children were shy and awkward and at a real disadvantage when confronted with the normal trappings of civilisation. Boarding school at twelve was almost inevitable for outback children and as a teacher I'd seen happy little extroverts from the bush become miserable and frightened, and often sullen and introverted, in an unfamiliar environment. It was bad enough to lose the wonderful freedom they had always known in the bush without the added discomforts of revealing to their taunting peers that such things as taps and toilets were an unknown quantity to them.

Bill had arranged the purchase of a six bedroom Shearing Quarters with the addition of a ten foot verandah all round from the Cyclone Company in Adelaide, as well as a combination workshop and store. We had no great faith in the Railways' ability to deliver a consignment intact to Alice Springs and would still need road transport the rest of the way, so decided to send two transports right to Adelaide to pick up the order at the factory and deliver it on site, so there'd be no off-loading along the way where pilfering might occur.

The Alice Springs Transport operators weren't about to fight over the job, lucrative though it might be. The prospect of driving their semis on a set of wheel tracks through hundreds of miles of sandy desert held no appeal and the only one who would consider it was Jack Litchfield, owner of a two-truck enterprise, who, with his other driver, Frank Fidler, had already made a delivery of horses and gear over the rough track.

Frank, now a man in his fifties, had already proved that the impossible could be done. As a young newly-married farmer in South Australia, Frank had been clearing his land and a tree had fallen across his back. The doctors said he would never walk again and, after months of arguing against his determined belief that he would, gave up on him and returned him to the care of his young wife, Oriel. Oriel watched as he lay in bed, gritted his teeth and concentrated on moving his legs. He did it every waking hour for many months and then came the day when, after an hour's struggle, he could sit on the side of the bed with his feet on the floor. The injury left him with permanently bowed legs and a curiously rolling, walking gait. The farm was gone; he moved to the Territory and took on a job which guaranteed exercise for his muscles, — driving transports, loading and unloading by hand, digging out of horrendous bogs. Pint-size Oriel went along with him whenever she could, clutching a thermos of tea and a knitting bag and lovingly admonishing him as "a silly old fool!"

Frank and Jack got the loading on site. Shortly afterwards Frank borrowed the money to buy a truck of his own as the expenses of raising four children were now over. When Jack retired due to illness, Joe said Frank should have all our business, even if it meant doing a turn-around with his one truck, when the road was better and the other operators vying for business. When cattle carting became common everyone wanted Frank, utterly reliable and nursing the cattle along as if every beast was his own. Instead of the long haul via The Granites we made a track across the forty-mile stage between Gangsters' bore and Refrigerator Well through the sandhills which bordered a large salt lake which we named Lake Fidler after Frank, the first to drive a semi-trailer that way.

For almost a year Joe had had no permanent headquarters, throwing out his swag each night and building his campfire wherever the job of the moment took him. Apart from supervising the track-riding of the cattle and cross-branding the cows and branding the calves with the new TST brand — on the flat without virtue of yards and with the help of those stockmen who could be spared from the Billiluna camp — he had also covered most of the 1620 square miles of country in his search for suitable bore sites to be investigated. He had cut timber to build the first set of yards and for the planned holding paddocks at each bore; organised the delivery of bore equipment, stores and some stud stallions, bulls, mares and cows which Bill Wilson had sent up from south; and made flying trips to Alice Springs to various government departments and to Halls Creek in search of urgent replacements for the unreliable white elements of his stock-camp workforce. Finding men he could rely on was not always easy.

Early in the year he had bought ten horses from Mt Allan station and in July he inspected and bought a hundred head from Indianna Station, north-east of Alice Springs. He gave the job of taking sixty of them overland to Mongrel Downs in early October to 21 year-old Malley Brown, the rest to be brought early the next year by

Dave Ledger, known as the Galloping Pom, along with his own horses en route for sale at Halls Creek.

Malley had grown up on Billiluna where his father had been manager for many years and he already had a reputation for fine horsemanship and good cattle nous. Malley was just the man Joe was looking for as head-stockman and was soon to become just like a member of the family, hero-worshipped by the children and a tower of strength to me whenever problems cropped up in Joe's absence. In addition, he had grown up as a child and worked in his teens with the Aboriginal lads on Billiluna and his particular friends— Harry Hall, Harry's cousin Rex and Bob Sturt — were to form the nucleus of our permanent stock-camp in the years to come. Of all our employees, Harry was the only one who was destined to stay with us for all our years on Mongrel Downs.

Joe's brother had a month's holiday in August and in that time he erected the homestead and workshop with the help of Aboriginal ringers turned builders' labourers. It was now September with the hot weather closing in, and a round of branding up the calves to be completed. I could not leave my High School classes so late in the year with a clear conscience, so it was decided that I see the year out at school and, with Malley to hold the fort and all the stockmen away on wet-season walkabout, it might be possible for Joe to take a holiday with the family before we started our big adventure together.

The week before Christmas I had an unexpected visitor. It was Lorna, from Finke, whom I had not seen for nearly five years, with her sons, Roger, two months older than Bobby, and Johnny, a curly-headed little half-caste boy about the same age as Tracey.

"I heard you and Joe goin' to start up a new station long way from here," she said. "I come to see if you want me work longa you again!"

I had vaguely realised that I couldn't look after that big house, do the bookwork, make the bread and teach the children without some help. Nothing could have suited me better. But everyone knew that you could not take a lone Aborigine away from his or her tribal country and relations and expect them to be happy. I tried to explain to Lorna just how isolated we would be.

"No matter!" she rejoined firmly. She told me that Mail Bob, her elderly tribal husband, had died cold weather time two years ago and the policeman had arranged for her to get a widow's pension to support herself and the boys.

Without Mail Bob or her eldest son, Ian, now working on a South Australian station, she was terrified for the safety of the two younger boys and herself. The referendum which had given Aborigines equal rights as citizens had also given them the right to drink alcohol and all those not actively employed now received a dole cheque. On pension day the Finke pub did a roaring trade with a black clientele, the publican unable to refuse them service, no matter how drunk they became, without risking the notoriety of a legal charge of racism. When their money ran out, some of the drinkers would soon think of Lorna and her pension money. Her camp was a mere two hundred yards from the pub. Every cheque had been taken from her and most times she had been bashed and sometimes raped too because the second and third groups of visitors would not believe that she had already handed over her money to the first arrivals.

"Sooner or later they kill me, Missus, and what my boys do then?"

I knew she wasn't exaggerating because in the first year of their new status at least eighty Aboriginal women had been murdered in the Alice Springs district in drunken brawls. My friend, the matron of the hospital, had told me that two or three died every week-end and nobody could do a thing about it because of the racist accusations from the southern media.

We had a cup of tea and worked out dates for Lorna to return to Alice Springs to meet us after our holiday. She expected the holiday time would be quieter when most of her people would go walkabout but she herself would stay in Finke; she had no intention of missing us. I said I would send her a telegram in good time for her to catch the weekly train to Alice Springs.

Luckily the office staff of the correspondence school in Adelaide worked over the holidays. I was able to arrange for two more sets of correspondence lessons for Lorna's boys to be included with those I had already ordered for Bobby and Tracey. The School agreed to send me sets for a whole term's work as we would have no regular mail service. Eddie Connellan only had single-engine planes on his mail runs and Mongrel Downs was considered "tiger country", too dangerous to service without a twin-engine plane. I could post the completed sets with passing travellers or, if none happened along, every three months when I went into town to pick up Kim for her school holidays. When I took her in to return to school the next term's work and a supply of children's library books would be waiting for me at the Post Office. (This arrangement continued for the full period of the children's primary education. It wasn't a problem. We had the daily School of the Air lessons with Nance Barrett and her successors to keep us in touch with any academic innovations.)

Kim had won a scholarship to a boarding school in Perth, half a continent away and we wanted to take her there and see the school and meet the teachers. The idea of a family holiday before we began our station life in earnest appealed to me as I knew it might well be the last we would ever have together. (As indeed it was.) Bill Wilson offered us his family seaside shack on Spencer's Gulf for a week before we flew across to Perth with Kim and Bob Sprod offered to put a car at our disposal. However, Joe had hired a carpenter to finish off the work on the homestead that Ken had not had time to do and Joe wasn't prepared to leave until it was completed. He took Kim as offsider and she spent the school holidays on the station accompanying Joe on the bore runs, riding her beloved Pi, and lamenting the fact that the demands of her education would prevent her being with us full-time.

One of the toughest years in Joe's life was now winding down. The cattle had settled and established their beats centring on the various bores and the new generation of calves, to whom the wide plains of Mongrel Downs was home, were wearing the TST brand and the bottletop earmark.

The plant horses had been turned out and the ringers returned to Billiluna to join their families in the annual Big Sunday walkabout. The Indiana stallion and the two Arab stallions were with their mares running on Andesite and Toyota Plains. Until the general rains of the wet season set in, the main job would be a regular bore run to ensure that all the bores were pumping and the supply of water constant for the cattle during the hottest period of the year.

Joe was ready for a break. Malley and Ferdy could be depended on to do the bore runs or pull a bore if necessary while he was away. Then the rumour circulating across the border finally reached him. The whole Halls Creek district was buzzing with the story that Joe had picked up a mob of Ruby Plains cattle and taken them across country and over the border onto Mongrel Downs. The newly-appointed Halls Creek stock inspector had seen them watering at the Lake Alec bore.

"Hell's bells!" exclaimed Joe, "I'd need to be a better man than Harry Readford to take cattle that distance across waterless country. What do they reckon I am! We'll get to the bottom of this!"

It was Christmas Eve. With Kim and Malley along for witnesses, he drove to Balgo. Some harsh words ensued but it soon became evident that, although they had passed on the rumour, it had not originated there. Joe established that the Stock Inspector had never been to Balgo or passed through, so he could not conceivably have been anywhere near Lake Alec. Then he drove north to Billiluna, where they spent Christmas Day in convivial company. Bill Moyle, owner of Karanya to the north, called in with his wife. Bill had heard the story direct from Mick Quilty at Ruby Plains.

The next stop was Ruby Plains. Mick was none too pleased to see Joe and said that the stock inspector had given him the story. It was after dark when Joe got to Halls Creek and the stock inspector was at a party. Joe pulled him out and confronted him. He was very abject and admitted he'd never been across the Border or anywhere near Mongrel Downs; actually it was Mick who'd complained to him; he was new to the district; believed what he'd been told.

Joe wasn't about to forgive him for passing the story around. He insisted that the stock inspector accompany him the next day back to Ruby Plains. Mick wasn't there; he was over at Karanya. So to Karanya they went. In the final confrontation Mick and the stock inspector accused each other and Bill Moyle either corroborated or disagreed with their statements. Joe let them argue and then gained the admission that nobody had seen Ruby Plains cattle on Mongrel Downs. "You're a boy in a man's job," he told the stock inspector, "and when I get to Perth I'll suggest your bosses transfer you to a job you can handle." (He did complain to the Lands and Agricultural Director and the Stock Inspector was subsequently transferred.) "As for the rest of the Kimberleys gossip-mongers, if I hear this rotten lie once again, I'll track it back to where it started and deal with it my way! You'd better put the record straight, Mick!" Then he walked out, satisfied that his good reputation was again intact.

Joe absolutely despised gossip; he'd seen too many people lumbered with unjust reputations because of malicious rumours and, if anyone ever made an unfounded comment to him, his reply was always "Come on, we'll go and ask him (the victim) if it's true. You said it; he's a right to know what you're saying and give you his side of the story!" The gossip-purveyor invariably backed down.

Our holiday together was great, except for the sadness of parting from Kim at the end. We handled the alien city culture pretty well for a bunch of bushies, though people did stare a bit when Bobby and Tracey left the plane carrying their cabin bags on their heads the way the Aborigines carried their gear at home, and I think Bob Sprod dined out on the story of Joe's efforts to return the car to the Adelaide business premises.

Nobody could lose Joe in a thousand square miles of bush country but the Adelaide streets had him slewed. He pulled up near a phone box at an intersection, phoned Bob, gave him a description of the surrounding buildings and asked for instructions.

"You're nearly there," said Bob. "Just drive another six streets south and you'll see the display rooms on the corner."

"How the hell do I tell which way is south?" burst out Joe. "I can't see the sun! The sky is completely overcast and I can't even see much of that for all the buildings blocking it out!"

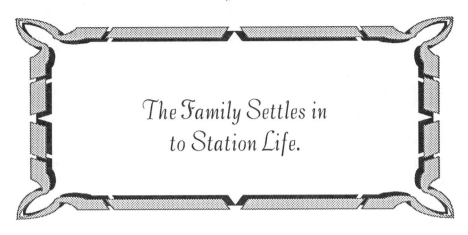

Chapter 11

The Family Settles in to Station Life.

What is the standard rule for the rest of the mob does not always apply to those who dwell in the outback, where a calculated risk is often a way of life. A one-ton vehicle can carry two tons at a pinch, as Gerry Adamson from Billiluna regularly demonstrated, and the notice on the turnoff 16 miles north of Alice Springs, stating that the road was for four wheel drive vehicles only, didn't apply to us. Joe reasoned that as long as we tackled the sandhills near Lake Fidler after sundown or before sun-up, when the cold night temperatures contracted and firmed the loose sand, he could drive our five-year-old station wagon right to Mongrel Downs.

Frank had loaded our household gear and the furniture I'd bought for the station on his truck; I'd arranged a tenant for our town house with the sleep-out retained for our own use; Lorna and the boys had arrived from Finke and at long last we were on our way. At the last moment I'd been persuaded to give Heather, a tall, beautiful English girl, a job as governess. I didn't have time to contact her last employer on a Barkley Tableland station for a reference; with no phones it would have taken a fortnight by mail. Heather had a lovely complexion and a beautiful speaking voice and was obviously intelligent even if her general aspect was a bit grubby.

It wasn't a trip I'd want to repeat but I didn't have time to dwell on it because when we got to the homestead we found that, with visitors and staff, there were twenty-six people lining up to be fed and I was the cook.

By the third day most of them had gone and Joe had left with the stock-camp. I had time to check on the verandah schoolroom where I'd set up the desks and given Heather the carton of school sets. She claimed to have been a governess on two stations so, I'd assumed she'd be competent. She was still wearing the clothes she'd arrived in, and slept in, and rather than supervise the laborious task of making the little ones print their letters for themselves she was filling in the sets herself and transferring plenty of grime to the pages as well. I hit the roof and we swapped jobs around. I didn't want a dirty cook, so Lorna and I shared that job and we gave Heather the laundry and unpacking that still remained to do, while I took over the teaching.

On the fifth day the Galloping Pom and Lollipop called in on their way to Halls Creek. Heather asked if she could have time off to go with them; they'd be back within a week, and I agreed. I thought she'd have a clean-up before she left, but she didn't. When Lorna went to burn off the rubbish in the drum she found a dinner plate from the new service still in its packing and Heather had already broken two glasses. I was too cowardly to sack her myself on such a short trial, but I was already aware that she was a lot more interested in sampling the variety of outback life rather than sticking to one job and doing it properly.

When she came back, still in the same dirty clothes and oblivious of my hints that the kids had finished with the bathroom, I hit on the solution. The stock-camp didn't have a regular cook; maybe she'd fit in there. I'd ask Joe when he next came back to the homestead for tucker supplies.

As it happened, he didn't come in but called me on the transceiver and asked me to make a batch of bread and get the dry rations from the store and bring them to the camp at Refrigerator Well. They'd had trouble with fights between some of the men and he didn't want to leave while the tension was high. I left on the Saturday afternoon, leaving the children in Heather's care for the twenty-four hours I'd be away.

Our stock-camp was helping to hold a Billiluna mob for handover to a drover and the row had developed in their camp with both whites and blacks involved and the cattle had rushed, so Joe wasn't very happy. Neither was I when I got home. A strange truck was parked by the shed. The kids met me bawling and Lorna followed them carrying little Jimmy and wearing a brow like a thundercloud. Two men had arrived in the truck from Halls Creek and Heather had locked the kids out of the house at nine o'clock that morning and they hadn't had any dinner or a bath or anything. It was just on sundown.

The revellers in the house hadn't heard me arrive. I had to thump on the door and bellow for entrance myself. Finally Heather heard me and opened the door. In the kitchen, bottles and glasses littered the table and the three of them were as drunk as monkeys. I did my block. I smashed the glass out of the hand one oaf was offering me and said coldly, "I'll give you quarter of an hour to get off the place. Get your things and go with them, Heather. They can drop you off at Refrigerator. Joe needs a cook, if you want the job, but don't come back here again!". They went.

In a few weeks we settled to a routine and it seemed as if we'd always been there, all our previous life encapsulated into a tiny closed fist and the present an open palm inviting adventure and romance, vibrant and alive. The outside world only touched us at tangents when I switched on the transceiver to send or receive telegrams or tune in to the School of the Air lessons. If anything else was happening in that other world, we were all too busy to take heed.

At the homestead Lorna did the housework and laundry and minded Jimmy while I was cook, teacher and book-keeper. School was a challenge and a delight. Bobby was in grade four and could work more or less by himself; Roger, in grade two, had to be pushed; Tracey and Johnny at five had open little minds keen to absorb the new skills of reading and writing and numbering. We started early and worked till lunch-time. I always finished by reading them a chapter from a children's classic or, if I was interrupted by visitors or minor demands on the Flying Doctor Medical Chest, I handed out butchers' paper and Craypas sets or plasticine and left them to it. I never interfered with their creative art experiments, just came back and admired when they had finished, but I was a tyrant about neat writing and correct spelling. After school they explored with Lorna, absorbing skills in bushcraft that were to make them far more in tune with the land than I could ever hope to become.

The horse muster and branding and breaking in the colts at the homestead yards was a time of excitement. I was tied to the kitchen and huge meals and smokos, but I was on the top rail when Malley, now head-stockman, told Bobby to choose himself a colt. The one he chose fought Malley every inch of the way when he was breaking it in, tried to savage him and ripped a sleeve out of his shirt with its teeth. Finally Malley pronounced it ready to mount. "Get your saddle, Bobby!" he called.

I saw my son's face go white, but he obediently climbed through the rails and walked towards the snorting, eye-rolling animal Malley was holding firmly. He looked so tiny standing with his saddle in the dust and sweat among the men and horses. Malley looked him up and down. "Y'know, Bobby," he said, "I reckon that colt over there would suit you better; he's got a lot softer mouth than this fella."

Bobby didn't need persuading. Malley legged him up on the substitute colt and he rode round and round the yard with a beatific grin on his face. He named the colt Nugget. For a fleeting moment I thought back to all those years of my own childhood when to own a horse of my own was the unattainable goal I had yearned for so desperately.

Nugget had to go into the horse plant with the other colts to learn his trade as a stock-horse, so then commenced an arrangement which endured until Bobby went away to boarding school. As soon as he got a fortnight ahead of schedule with his school sets, he was allowed to go out in the stock-camp for the next two weeks.

Joe put up a one-room building for Lorna and, much to the ringers' disgust, netted in the big bough shed for a chookhouse for me, while the kids chased and captured the chooks which had run wild and laid eggs indiscriminately in the surrounding scrub for nearly a year. He fenced a garden area and laid on the water and planted a citrus orchard close by. In the workshop a big goanna named Mechanic watched over the men's shoulders as they worked at bench or forge and a smaller one named Apprentice hovered in the background. With the approach of the bitter cold of the Centralian winter Joe made a "donkey" of two forty-four gallon drums over a fire-box and connected pipes to the house so we had hot water in kitchen, laundry and bathroom.

My job was to paint the homestead rooms, one undercoat and two coats of paint. I did it on alternate week-ends to the book-keeping. (It took me nearly two years.)

Kim was due home in May for school holidays. Swags, water, fuel, tuckerbox were loaded in the back of the station-wagon. At the last moment Joe added the welder needing repairs he couldn't handle. It was 120 miles to Chiller Well and the Lake Fidler sandhills were about mid-way between. I took the children and called in on the way with rations for the stock-camp near Madame Pele's hills, where Bobby elected to stay with Malley until I picked him up on the way home. The track through the sandhills was silver in the moonlight, the sand-grains contracted to an almost-deceptive firmness. We camped the rest of the night at Refrigerator Well and in the morning I rigged an aerial over a tree branch and called Joe to let him know I'd survived the sandhills and wouldn't need him to come and rescue me.

When I called at Chiller Well, I found Margaret (Lollipop) alone. She told me how Heather had disrupted the cattle camp at Refrigerator by being discovered too often stripped to the waist having a wash when the men returned to camp. I remarked that this washing was right out of character according to my experience. "Different circumstances," laughed Margaret. "Dave felt sorry for Heather when Joe sacked her and said she could keep me company here for a while. She was keen enough on her daily shower while he was here!"

The Chiller homestead was only a two-room, dirt-floored shed with a tarp around four posts out the back for a bathroom. Twice in the week while Dave (the Galloping Pom) was home, Heather had screamed that there was a snake crawling under the tarp. Margaret and Dave had been together less than a year. The day after Dave left with his stock-camp, a stock inspector on the way to inspect the drover's mob called in and Margaret got Heather, despite her reluctance, a ride back to Alice Springs, from where she disappeared from our ken.

I thought I'd be in Alice Springs by nightfall but the calculated risk of adding the welder to the load hadn't taken account of the position of the welder, right over the back wheels. South of Yuendemu the near-side back wheel buckled. I put on the spare and crossed my fingers. The off-side wheel collapsed just near the Hamilton Downs turn-off.

I was just about to make camp, when the Hamilton Downs manager came along and took us in to the station for the night. In those days you could leave a vehicle broken down on the road without any fear of theft. The next day he lent me two spare wheels and the rest of the trip was a breeze.

I did the shopping, collected the mail, replaced the tyres, met Kim's plane and left the welder in town to come out with a later loading. Kim didn't want to waste a minute of her holidays in town, so we left early, delivered mail and bread and had tea with Margaret at Chiller Well. We then drove on to camp the night at Refrigerator. We tackled the sandhills before sun-up and picked up Bobby at the camp near our boundary at smoko time and were home for a late lunch. I was pretty smug about my effort but what I didn't realise then was that I'd ruined my chances of any leisurely trips to town in the passenger seat beside my husband. Joe was always going to be too busy to do a trip he could delegate and that's why, after the blue-grey sedan replaced the too often overloaded station wagon, my regular trips to town earned me the nickname of *The Blue Flyer*.

One afternoon in late August 1965 I was alone in the house when Lorna came running to me, weeping and distraught.

"Missus, Missus, brother belonga me close-up finish! He longa Alice Springs. You take me, please, you take me!"

She was hysterical. All I could get out of her was that a blue jay, which was her brother's dreaming, had alighted on her verandah bed where she was resting, told her that her brother was deathly sick and then collapsed and died itself.

I had already planned to go to Alice Springs in two days time to meet Kim for her holidays and Joe had come in from the stock-camp the night before to be at the homestead while I was away. The state Lorna was in was such that if I didn't take her to town right away she would probably start walking. Joe said he'd see that the kids finished their schoolwork sets and Johnny and Roger could sleep on our verandah with our kids instead of alone in their little house.

Within half an hour Lorna and I were away. We camped that night at Refrigerator Well and got into Alice Springs at mid-afternoon the next day. Lorna's brother lived at Ernabella, over the border in South Australia, but she was adamant that he was in Alice Springs. When she jumped out of the vehicle and left me, I hoped she would soon find some Aboriginal acquaintances who could re-assure her that her brother was all right.

She came back to our town house at tea-time. Her brother had been flown in by aerial ambulance from Ernabella the previous afternoon but had died in the hospital soon after arrival. I well knew the power of mental telepathy between Aboriginal people but the death of the "dreaming" bird at the same time as the man died was truly an enigma.

Lorna stayed in town over the school holidays and when I picked her up to go home I sensed that our time together was nearly over. Mongrel Downs was just too lonely for her without the companionship of her own people. She agreed to stay until the boys finished the year's schooling and Johnny's white father, now living in Alice Springs, agreed to provide accommodation there for the little family and ensure that the boys went to school.

We were really sorry to see them go. The boys had been good mates for our children and I missed Lorna's company and the fact that she made it so easy for me to teach by minding three-year-old Jimmy during school hours.

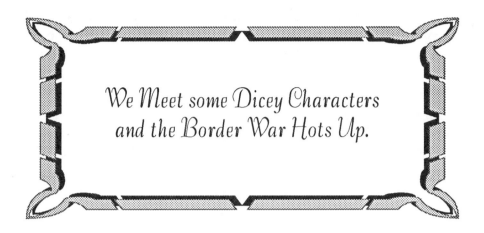

Chapter 12

We Meet some Dicey Characters
and the Border War Hots Up.

Horse trading in the Australian bush has always been fraught with interesting possibilities. This was doubly so when stockhorses in the Kimberleys were at a premium because so many succumbed each year to Kimberley Walkabout Disease. Balgo, south of the danger zone, bred horses for this market. In my first year at the station the Galloping Pom passed through with a mob of horses he hoped to sell to various Kimberley stations. He was pulled up with a jerk at the Balgo border. Father Mack, again with official police back-up, denied him right of passage on the grounds that his horses might introduce "the dreaded Birdsville Disease" to Western Australia. In vain did the Galloping Pom argue that this disease was only caused by horses eating the indigofera plant, none of which grew within hundreds of miles of where his horses had been pastured, and anyway, it wasn't a contagious disease but a plant poison, so it couldn't possibly be passed on. The authorities denied him a permit. He asked Joe if he could turn the horses loose on our side of the border and pick them up later when he had sorted out the problem and Joe agreed.

It was six months later before grudging permission was given to take the horses into WA. The Galloping Pom had missed that year's sales and Balgo had once again cornered the market. The Pom was philosophical about it: "You win some; you lose some. At least, I'll now have the horses right on hand for next year!"

Shortly after that event Joe decided to take a run to our northernmost new bore at MacFarlane's Hills to get a water sample for analysis before we put any cattle there. To his great surprise he found the cross-country tracks of a truck, not more than two days old and, on searching around, the tracks of three or four men and, where the truck had parked, a few particles of what was undoubtedly baled hay.

As far as he knew, the only people who knew of the existence of that bore were our ringers and Ferdy and Dangerous Dan. He decided to camp there out of interest to see just who was planning to use the watering point. The hay indicated that the stock would be horses and not cattle. He waited for two days but then had to make a flying trip home for other urgent demands. When he returned he had missed the travellers. A considerable mob of horses had watered at the trough and the tread of the truck tyres was clear in the soft sand.

A fortnight before we had heard on the radio that a mob of horses had been stolen from Victoria River Downs but we had decided that that was a bit too far north, even for the Black Hills boys, who'd been nicknamed the Two Jacks, (Hydraulic and Trewallah, — what one couldn't lift, the other would). Now we weren't so sure.

When the Black Hills boys were arrested some months later, tried and found guilty of the theft of the VRD horses, it was easy to draw conclusions about where the horses had ended up. Official police enquiries could not locate the horses anywhere in the Territory. The Black Hills boys admitted the theft and accepted the consequences, but they weren't about to break the long-standing creed of outback Territorians and involve anyone else. Had they planted the horses in some hidden gorge? Had they sold them? Their lips were sealed.

Joe had estimated the horse numbers by the tracks at the bore and they tallied with the number stolen. He had recognised the tread of the truck tyres as belonging to a Kimberley vehicle which had passed through our property on previous occasions. (It is second nature to a bushman to read and pinpoint vehicle tracks in the desert country through which so few vehicles travel.)

"The Galloping Pom is going to have a bit more competition than he expects in the next horse sale," he remarked to me.

When next he met the owner of the suspect truck he couldn't resist the comment,

"No shortage of horses for the Kimberley market coming up, I hear!"

All he got in reply was a sour glance.

The Black Hills Boys were two likeable young fellows trying to start off their block with no capital behind them, working on the principle that it was okay to lift things as long as you didn't lift them off another battler. They took contract jobs wherever they could and in our first year Bill Wilson had given them the job of digging a well in the bloodwood grove forty miles north-west of Refrigerator Well, a long stage for walking cattle but the only likely place where water might be found.

They weren't there when Bill and Joe arrived one day, but Bill elected to climb into the well to see how far down they'd got. Six feet below the surface a deep recess was cut into the side of the well and stacked in it were supplies of dry rations and tinned goods, the bags bearing the government stamp. The boys had come straight from a job on Hooker Creek Settlement, but nobody was going to find a stocked larder in their shack at the Black Hills if there was an investigation. When they'd finished the well and got a good water supply one of them named it Sangster's Well after his girlfriend, but Bill changed the name to Gangsters' and we never knew it as anything else.

Joe always claimed that Bill Wilson was the best he'd ever known when it came to reading tracks, on a par with the legendary Jim Campbell who was claimed to be so good that he could track the Holy Ghost through a thundercloud. Bill could glance at the track of a man or beast and say who was there, how long ago, and whether they were in a hurry or just taking their time.

From a few tracks at bore or roadside Bill would automatically reconstruct a scenario of events that had taken place there. This skill, to a greater or lesser extent, is possessed by all bushmen, but Bill could add the finer points.

It was at Gangsters' Well that Joe's expertise was called upon. He drove up one day to discover water pouring out through two bullet holes in the stock-tank. He plugged them quickly. With no water for forty miles either way, what criminal fool would do such a thing? He was livid with anger. There were eight shots in all, but all except two had been deflected off the metal sides of the tank. Another six or so were embedded in the bloodwood trunks behind the trough. A large truck had been parked beside the trough and its lone occupant had boiled the billy and had a meal. His footwear was not the elastic-sided boots worn by all bushmen but sandshoes with a distinctive pattern on the sole. Joe sketched the shoe size and the sole pattern. Then he checked the angles of the bullets and estimated where they had been fired from. He walked back along the road and sure enough the vandal had knelt down on the wind-row to take aim. It was the work of a moment to hunt around and pick up fourteen .22 cartridge cases. Each one had a tiny groove close to the rim.

He knew the culprit must belong to the Government Mapping team which had been camped not far from the homestead for the past fortnight, and some of the personnel were straight from the southern cities. Only an outsider could be stupid enough not to know that water at a given point could mean the difference between life and death in the bush.

Joe came home and showed me the evidence and then drove to the camp and showed it to the chief. The young fellow who had driven the truck from town three days earlier denied owning a gun.

The owner of the only gun in the camp agreed to fire off a round, and there was the tell-tale mark on the cartridge case. Then he remembered that he had left his gun in the truck before it went to town. The only sandshoes with the sole pattern corresponding to Joe's sketch belonged to the young driver. His boss sacked him on the spot and apologised to Joe profusely, and the next day took the offender back to Alice Springs. This was one of the few occasions where a vandal of this type was caught; unfortunately, as tourism developed, more gun-happy fools found their way into regions which had previously been spared their attentions.

Water in that arid land is the most precious commodity of all and we were twice involved in the rescue of travellers who would otherwise have perished. Ferdy found the first pair broken down between Granites and Tanami. They had drunk the water out of the radiator and even a bottle of vinegar from their tuckerbox. They had written out their wills and were on their last legs. When they had drained his waterbag and dried their tears of relief, he towed their vehicle to the homestead and we repaired it for them and made them welcome. They showed us their map — a page torn out of a school atlas — and said they had expected to get fuel and supplies at Granites. At this time both Granites and Tanami had been long abandoned. Joe gave them enough fuel to get to Balgo and sent them on their way after a couple of days when I complained that they were making themselves a bit too free with our tools and stores. We found out later that they were wanted by the Alice Springs police.

Within six months another pair of wanted men fled in our direction. It was midsummer and Joe was doing the bore run and found three-day-old vehicle tracks at Ferdy's bore. They were obviously out of fuel because they had taken diesel fuel from the drum left there for the pumpjack and must have been hungry too because they had killed and butchered a calf. Joe knew they wouldn't get far on the diesel fuel and, annoyed though he was, he set out to follow them. He found them thirty miles along the road, starving again and very sorry for themselves. He fed and refuelled them. They had no money but Joe forbore to lecture them because he knew he could rely on Father Mack to do that when they made their unwelcome appearance at Balgo. When we learned that they too were fugitives from justice, it occurred to me that such characters might easily have decided to attack Joe to replace their stolen vehicle with his. He laughed at my fears but I was reassured when he fitted a strap on the inside of the driver's side door to carry a holstered pistol. Maybe the Territory was no longer frontier country, with improved communications and many amenities we had never known before, but the outback was also a more dangerous place than it had previously been.

We weren't keen on seeing our road become a highway to the west for all comers, so were pleased when the government graded a good road linking Tanami to the border and the Halls Creek road, thus bypassing both Mongrel Downs and Balgo. Our track was shorter in distance but the other road north of us was a straight formed road, much easier to drive on.

Then the nuisance value of the fun we had extracted from directing unwelcome travellers through Balgo rebounded on us. Father Mack reasoned that there would always be some who would opt for the shorter distance and a road joining the north road to our MacFarlanes bore would be shorter still, turning off many miles north of Balgo but passing within two miles of our homestead. His only problem was that his planned connection was in the Territory, not Western Australia. Somehow he managed to persuade the Halls Creek roadworks team to do the job and even hired the Balgo grader out to them. He couldn't have the poor nuns terrorised by no-hopers, he said. I thought that was a bit rough, seeing I was more often than not alone on the station with the kids, while the nuns had the back-up of school-teachers, the male clergy, nursing staff, mechanics and a couple of hundred Aborigines.

We didn't know a thing about the project until one morning Joe went to check MacFarlanes bore and heard in the distance what could only be heavy machinery. He thought he was dreaming. He drove across-country towards the sound and could hardly believe his eyes when he saw a surveyed line and, in the distance, a grader working on a formed road, a road camp and two roadworks utilities. They were within five miles of the bore.

It took a lot to rouse Joe, but when he did blow, nobody ever forgot it. He blew then. All work ceased. The Western Australian government official in charge of the job seemed to believe that an agreement had been reached with the Territory government.

Father Mack and his boss knew all about it; he was puzzled that Joe didn't. Joe advised the Balgo grader driver in some detail what he would do to the grader if it wasn't off our country in the next half hour, but he allowed that the surveyor and his team would need a couple of hours seeing they had to dismantle their camp. He parked his vehicle across the end of the road and supervised the withdrawal. When the last utility disappeared in a cloud of dust, he slipped home fast and sent a wordy telegram to Alice Springs.

Two days later two inspectors arrived and were shown the offending road; they said they wouldn't have believed it, unless they'd actually seen it. They assured us that no way would they sanction its completion when there was a perfectly good road to Tanami. I fed them the best steak and put on a fairly convincing act of the lone little woman terrified at the prospect of fending off God-Knows-What sort of malcontents on evil bent. It was bad enough with the few that did come our way without a highway past the door bringing them in droves. They left to send some stern paper work to Halls Creek and we heard later that Halls Creek refused to pay Balgo for the grader hire.

It was a close shave because once the road was there it would certainly have been used. It wasn't so much that I was afraid of unpleasant visitors but that, when you are the only habitation in hundreds of miles, you find you are constantly interrupted by travellers wanting fuel or breakdown assistance or just calling out of curiosity and asking damnfool questions. Three or four of those a day and increasing busloads of trippers and you may as well set up a roadhouse to cater for them. I had neither the time nor the inclination to go into the roadhouse business.

Private exploration and initiative were responsible for making the road from Refrigerator Well, through Mongrel Downs and right to Balgo Mission in Western Australia, but it wasn't long before various government departments wanted a piece of the action.

The first missive was from the NT Aboriginal Welfare Department to the effect that we couldn't drive on the road through Yuendemu Reserve without a permit each time for each person, and to get the permit we were required to have a prior medical examination at Alice Springs hospital, showing us free of any germs that we might conceivably pass on to the Yuendemu Aborigines. Rather difficult, this one, because we would have to go through Yuendemu first to get to Alice Springs to get the permit to go through Yuendemu. We ignored that one. The road we used didn't go within miles of the settlement, anyway, so we rarely saw any of the people, except for an occasional group beside a broken-down vehicle returning from town, where the germy town population could have breathed on them at will. Being Aborigines, they didn't need permits or medicals.

The next advice came from the NT Wildlife Department, telling us that we would need a permit from Darwin each time we used our road through the Tanami Wildlife Reserve, and that firearms could not be taken onto the Reserve under any circumstances. There weren't any signs anywhere to show an approximate border for the Reserve where, as law-abiding citizens, we would have to chuck out any firearms from our vehicles. "Forbidden to use" would have made a lot more sense than "Forbidden to carry"! We ignored that one too.

We also ignored the one from the Western Australian Aboriginal Affairs Department, telling us that we couldn't go over to Balgo or Billiluna without first getting a permit from Perth. Consider that we didn't have a mail service, only irregular mailbags sometimes six weeks or more apart. How were we supposed to apply for and get back these permits from Alice Springs, Darwin and Perth for a journey necessary in the next day or so? What earthly difference would it make whether all, some, or none of us carried a piece of paper with us on a desert road where two vehicles a week was the exception?

These red-tape artists, these clerks in the woodwork of government obfuscation, these paper-pushing, nuisance-value twits were asking the impossible, unless they, all three, would decide to give us and all our employees an open permit to travel on our own home-made road. Which they wouldn't, of course! What happened? Well, as far as any of those silvertail clerks ever knew, no one on Mongrel Downs ever left the station, no one ever visited, or transported cattle, or brought loading, or even passed through on the way to the Kimberleys. That way, everyone was happy!

Mahoods of the Outback

·········· indicates travel to new homesteads in 1962 and 1972

Chapter 13

Good Seasons and Good Markets.
The Exclusive Tanami Yacht Club.

The wet season of early 1966 was a bumper. As there were no rivers to carry away the rain, it all stayed on the property, accumulated in swamps and the fresh-water lakes — Bullock's Head and the four named after the Wilson children, Ruth, Alec, Emma and Sarah. Lake Ruth was a mile across and twenty feet deep in the middle and set to last for years.

The cattle were literally knee-deep in grass on the plains and in soft spinifex on the ridges and the lone old buffalo, whose tracks had first been seen on the return expedition, had concluded a long exile and joined a mob of cattle near Lake Alec. There were limitless carpets of wildflowers and such a variety of birds that the bird book was referred to almost daily. Joe always carried a framework for pressing plants in his vehicle and he discovered two Australian firsts, which had to be sent to Kew Gardens to be named, and three Territory firsts. The children found apus (those strange trilobite-looking creatures known in the desert as "claypan tadpoles"), in Lake Ruth and tiny, nutty-flavoured onion-like bulbs buried in the sandy shore, which they roasted and stuffed in their pockets to eat whenever the fancy took them. Nature was so prolific that each day brought a new magic to observe.

We were in the right place at the right time. Not only was the weather on our side but the markets were also stable. The eleven-year Beef Agreement with England was still in place and a new American trade was emerging, so that cattle prices were steadily improving.

The pattern of our year was set. Horse mustering in March and the breaking-in of the colts. With Malley as head-stockman, the stock-camp comprised six Aboriginal ringers, all single boys from Billiluna, keen to see new country. (There was no attached native camp as on most other stations at that time because there were no local people nor any knowledge of the country among the Aborigines of surrounding districts.) Each ringer chose six horses for his string, the horses he would ride and care for during the muster, two of which were newly-broken colts to be trained.

The first round of mustering was to pick up the sale mob and brand the calves by the open-bronco method out on the flat. Timing was important because the sale cattle had to be in Alice Springs to go on the train booked for a certain date. For the first couple of years our cattle were handed over to the Billiluna drover on the way through, but when the road improved we changed to road transport all the way. Joe coincided his delivery dates with the week before the Halls Creek races so the ringers, after the gruelling work of the past months, could take a week's well-earned break. Joe, ahead of his time, always paid the full award wages to the Aboriginal ringers and I kept their store accounts and did their tax for them and persuaded some to open savings bank accounts. They soon learned not to draw all their money for the races because, by tribal law, whatever they had in the way of cash or clothes could be, and usually was, demanded by senior tribal relatives. They always returned in ragged clothes, barefooted and without their swags and had to be re-outfitted from the store, with the exception of pint-size, young Scotty Surprise, who packed away his good clothes and belongings, handed them to me for safe-keeping and left the station, bootless and hatless, in his oldest shirt and trousers.

Scotty was the only one who didn't smoke and his savings grew steadily. Scotty withstood all temptations to show off to the girls in flash gear and delighted us on his return with graphic descriptions of the races, the rodeo rides and the amorous adventures of his mates. His own adventures were no less absorbing. He came from Fitzroy Crossing country and, because he was so small, he had not been circumcised in the tribal ritual with the boys of his own age and the tribal elders were now insistent that it should be done as soon as they could catch up with him. He made quick sorties to visit his mother and take her presents and had so far evaded capture by meekly agreeing to the ceremony when he arrived and then, while preparations were in hand, slipping away to buy a plane ticket with the money he always secreted for a necessary quick getaway from The Crossing to Halls Creek.

The second round of mustering and branding was a more leisurely affair and the year culminated with a Christmas tree and Christmas dinner for everyone at the homestead and games on the lawn. Joe always offered the Aborigines employment through the Wet to help with fencing and yard-building but Long Johnny was the only one who ever stayed right through. Southern media claims that exploiting Territory pastoralists sacked their Aboriginal staff after the cattle work was over were simply not true; the Aborigines, just like anyone else, wanted a holiday or had private tribal business, and if they took three months off instead of three weeks, that was their choice and the employers had to fit in with it. By New Year the station population was usually reduced to just the family and the occasional visitor until the increasing storms closed the road until the Wet was over.

The roads to Territory homesteads were graded annually by a government grader and our road had originally been made from the Granites Road turnoff by Joe picking the track ahead in his vehicle and, at intervals, lighting a clump of spinifex to put up a smoke for the grader driver to aim for. Grading, the next year, was a long arduous job, working for days on end in almost-featureless country without a sole passer-by to relieve the monotony. The grader driver liked to relate the story of how he breasted one more sandridge, after the hundreds that had gone before, to see suddenly spread out before him the dry oval of Lake Ruth, on which a fully-fledged baseball game was taking place. He thought he was dreaming. His arrival had co-incided with Joe giving the men a couple of days off to wash their clothes and shoe their horses. When the Billiluna team arrived and the competitive skiting began, Joe suggested the match, produced a pick-axe handle for a bat and made a ball of tightly-wadded rags. What was lacking in finesse was compensated in enthusiasm.

As they neared Lake Ruth the following year, the grader driver told his new offsider of the incident, a "one-up", he said, you'd never be likely to strike anything like that again. They came over the rise. It was the September school holidays after the big Wet and we had visiting family and friends. A mile-long stretch of sparkling water greeted the grader-drivers' eyes. A small, white-sailed yacht scudded along with two ten-year old boys in yachting caps at the helm and two more swimming behind. On the white beach three teen-age girls in bikinis sunbaked and smaller children were building sandcastles and paddling on large red surfboards. Before they'd got their mouths properly closed, the kids invited them to become honorary members of the exclusive Tanami Yacht Club.

Despite the grading the road was frightfully slow because there was, beneath the sand cover, the undiscerned foundations of ancient, eroded anthills, ten or twelve feet in diameter, the fore-runners of those immense structures which still stud that landscape and are found nowhere else in Australia. With the passage of later vehicles the sand bedded down and the cement-like bases began to protrude here, there and everywhere. Ten miles an hour was a good speed if you didn't want to wreck your vehicle. The solution to our road problem came from an unexpected quarter. A big mining concern was being set up in the West Australian desert south of Balgo and the equipment was trucked, fifty tons to a load, across our road to Balgo and then down into the desert. The antbed foundations crumbled under repeated loads of that weight and were gradually spread so that eventually we could drive to town in a single day.

I had a variety of adventures on that road, especially as the station wagon grew more decrepit and cantankerous. When it broke down, I would rig the transceiver, wait for the "galah" session when the Base closed transmission for the day and the bush people chatted to each other, and explain my predicament to Joe, with a good proportion of the outback listening in. At first I didn't know the names of the bits that didn't appear to be working and referred to them as thingummys or whatsits, but I could usually rely on a disembodied voice chiming in with "I think she means the carby, Joe," or "Sounds like the alternator to me, Joe". I could change a wheel, clean the battery terminals, rope the exhaust pipe back on with fencing wire and dig out of a bog but I was always a bit shaky about what went on under the bonnet.

My worst experience, however, was when the kids opened a box containing white mice, which I didn't know they had, just as I was negotiating the tricky curves of the Lake Fidler sandhills. The mice leapt everywhere, but mainly at me. I screamed and let go the wheel and I think I would have jumped out if the door hadn't been roped closed because the lock had ceased to function. Honestly, the station wagon drove itself that time because we didn't get bogged and when we finally came to a shaky stop past the danger stretch the kids burst out laughing. My upswept salon hair-do, which I'd been nursing home to show off to my darling husband, had suffered badly in the clawing and beating I'd been giving it and one lone mouse was swinging on a long, disengaged strand below my left ear. Tracey removed it tenderly and accused me of frightening it. I wouldn't get back in the car until they assured me they'd caught the lot, which they hadn't, but luckily I didn't know that till later.

My experience with the drunks wasn't so funny in retrospect. It was well after dark and I was driving the Landcruiser with the two little ones in front with me and Kim and Bobby lying on their swags on the back. My headlights picked up a car jacked up with a wheel off and a campfire lit by the roadside and a single figure waving me down. Travellers always helped me when I was in trouble so I automatically slowed down. On a lonely road like this a breakdown could spell disaster if no help came along, and it would normally have been unthinkable not to stop.

It all happened so quickly. The figure reeled to my door and a blast of rum-scented breath hit me in the face. Other shadowy figures were moving from behind the car. When he registered that only a woman and children were in our vehicle, the expression changed to a wolfish leer and if ever I saw "Rape!" writ large and clear on a face it was then. He made a grab at the ignition key. I snatched up the heavy wrench lying beside my seat and smacked him across the face as hard as I could. He fell back screaming.

"Are you kids all right in the back, Kim?" I yelled through my window.

"Yes, yes," she yelled back.

I put my foot down on the accelerator and, despite my shaking knees, I didn't pull up to camp for the night until I'd put another fifty miles on the gauge. Luckily Kim had been wearing riding boots and when the other Aborigines had tried to climb on the back, she'd hammered their reaching hands with the heels of her boots. "A dog would have been handy, though," she said.

After that, Joe told me never to pull up for broken-down Aboriginal vehicles. The turn-off to Yuendemu settlement was on our road home and the Aborigines frequently left town with a load of grog but not enough petrol to get home. The police advised me to carry a pistol and helped me choose one at the gun shop. Then one of our station-owner acquaintances, an old man living alone, who had always been a friend of the Aborigines, was murdered by a group, to whom he had already given a spare wheel from his workshop, when he refused to also put it on for them.

It was a sad state of affairs. The grog was a problem which would destroy many decent Aboriginal lives and the catalyst which triggered distrust and fear between black and white to an unprecedented extent.

We didn't have any trouble at Mongrel Downs because we followed the example of many other properties and made a No Grog rule for everyone but Joe could no longer give our Aboriginal Stockmen the station vehicle to go to the Halls Creek races by themselves. It wasn't entirely that we didn't trust them, but the responsibility of keeping the vehicle safe from other drunken joy-riders was too much to expect of them.

Taking our staff to Alice Springs took on a new aspect. Our unsophisticated bush Aborigines, easily recognisable in tight pants, big hats and high-heeled boots, and obviously with money in their pockets, had only to walk up the street to collect a crowd of hangers-on urging them towards the pub. Their first purchases were always new clothes and, after a few experiences of theft and even being stripped when they drank too much and passed out themselves, they hit on the idea of making their purchases and leaving most of them in the shop for Joe, Kim or me to pick up later.

Often they landed in jail, sometimes because they were innocently involved in a brawl in the street or because they had refused to hand over their money or shout for the town gangs. Scotty Surprise was too smart for the town hangers-on though; he bought all his gear by mail order and, being so small, he could pass for a boy and thus avoid their notice.

It became almost routine, when leaving town, to go round to the Police Station to bail out any missing staff members. The bail was set at the value of the usual fine, so we waived the Court appearances and left straight away. Once, when I had taken Kim in to catch her plane back to school, Long Johnny had accompanied me to help load up the Landcruiser. He'd camped in our garage and stayed out of trouble for three days and I told him to be back at the house to leave at five o'clock. At six I went looking for him. A group outside the pub cheerfully told me, "P'leece bin takim, Missus! He bin proper drunk and fightin.'"

I was mad. It was now too late to leave and I didn't fancy a still-drunk companion; I'd let him stew there all night. When I turned up at six the next morning to bail him out, the young policeman burst out laughing.

"Are you his boss?" he chuckled, sizing up my five feet two. "He's been bellowing all night — Where's my boss? My boss can't leave me! — If he wasn't so big, I reckon the others would have thumped him; nobody got any sleep."

I paid up and he came out, dragging his feet and hanging his head. When he went to get in the passenger seat, I snarled, "In the back. I don't want you throwing up in the front!" I didn't relent until we were three hours out of town and I pulled up for a drink out of the waterbag. He took a long drink and climbed in meekly. He was very quiet for a while. Then he said, "You gonna tell Joe, Missus?".

I thought what a reliable offsider he was to Joe and how genuinely good he was with the kids. I remembered with a grin how he had taught Bobby to gamble blackfellow fashion and Bobby came home proudly with his winnings - a pocket knife and a pair of Long Johnny's socks.

"Not this time, I guess, Johnny," I said. He brightened considerably and gave me a spell with the driving.

Chapter 14

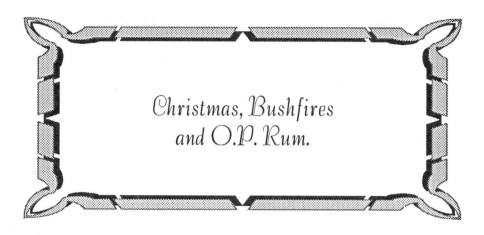

Christmas, Bushfires
and O.P. Rum.

A week before Christmas 1967, Brother Michael called in the Mission truck, had the obligatory cup of tea and took the mail to post. He dourly agreed to pick up our mailbag from the stock agents and a bag of bread from the baker. I suspect the good Brother only called in because the fiery Father had ordered him to; I never once saw him smile and he always gave me the impression that he felt the Lord had let him down. Maybe He did too, because when the truck pulled in three days before Christmas, Brother Michael was in a foul mood. He refused smoko, grunted that his five Aboriginal passengers had all got on the grog in town and he'd had the devil of a job rounding them up. He'd had to load up by himself, had picked up the bread but, once he had his reluctant passengers on the truck, he couldn't take the chance of stopping at the agents for the mail for fear they all jumped off and disappeared again. His muttering passengers slung down the bag of bread and the muttering driver clashed the gears and drove off.

My six bars of bread, warm from the baker's, had been sat on for 400 miles and squashed flat. Bobby took the bag straight over to the chook-house and tipped it out. I sat on the back step glowering.

"I *am* going to have the mail and some town bread for Christmas!" I announced suddenly to the kids. "I'll go myself!"

"You can't, Mum," said Kim. "Dad's got the portable transceiver, and he won't be home till Christmas Eve!"

"I'll go without it this time," I said firmly. "I'll get as far as Chiller late tonight, and give you a call in the morning from there on Margaret's transceiver. If I don't call in, you'd better tell Dad when he calls you, but I'll make it all right. You can cook the chooks on Christmas Eve, Kim, like we always do, and the rest of you, Kim's the boss, give her a hand where you can, and, Trace, you can ice the cake. Bobby, you go and fuel up the car and check the water canteen while I grab something for the tucker box!"

I only had one thought in mind. We hadn't had a mailbag for weeks and I *was* going to have that special mail. By nine o'clock that night I was smugly congratulating myself that I'd got through the sandhills okay and only thirty miles to go to Chiller Well, when suddenly my headlights went dead. Luckily I was on a straight stretch of road. Any other time it wouldn't have worried me because I'd just have camped until morning, but if I didn't get to Chiller in time to call Kim she'd press the panic button and Joe would have to leave his job to come looking for me, when I didn't need rescuing, and what a fool I'd look. What to do? It wasn't moonlight. The windrow was sand two feet deep; I'd probably get bogged if I tried to drive in the starlight. In the end, I tied the Dolphin torch to the nearside bumper bar with my shoelace and the beam was just strong enough to pick out the windrow and keep me on track, but it was slow going and took me the best part of two hours to reach my goal. Margaret had been alone for a week; she expected Dave back anytime.

"Come to town with me," I said on impulse. "You can leave him a note and we could even be back before him."

"You're on!" she said.

We only waited in the morning to call Kim on the transceiver and then we were away. We got to town mid-afternoon, collected mail, bread and an unexpected bonus of various stone fruits, which had just come in on the train especially for the festive season. We put in a short appearance at a party and left town at first light next morning.

Margaret was delightful company so the long journey seemed to pass quickly and we were back at Chiller with time to spare. Dave still wasn't home. Margaret lit the fire and put the billy on and we lugged in the huge parcel she had collected from the Post Office.

"It's our wedding present from Mum and Dad," she said, fingering the English stamps nostalgically. She unpacked it carefully — a full Royal Doulton dinner service; there was hardly an inch to spare on the rough bush table when it was all unpacked.

We'll christen it," she said, and rinsed two cups, set them on their saucers and poured the tea.

"No cucumber sandwiches?" I remarked.

"No, but I'll cut my Christmas cake, that's better!"

We had hardly taken the first sip when from behind the cotton curtain which protected the shelves from the dust, where Margaret kept all her stores, there came a loud cheep. We stared at each other. Margaret pulled the curtain aside. On top of the bowl of eggs a yellow chicken slipped about and complained stridently. Three other eggs were cheeping through their cracks. The heat of the corrugated iron wall was as good as any mother hen.

"They wouldn't believe this at home," remarked Margaret. Her father was a professor at a London University and Christmas in the two-room, dirt floored bush homestead, in the midsummer heat, could not have been further removed from all those she had experienced in her parent's home in snowy England. Except for the chinaware; that was the only common touch.

Dave and I passed at the turnoff and yelled "Merry Christmas!" to each other. Four hours later I was home. Sounds of corroboree from the ringers' camp; Joe and Mallee and Ferdy having a drink on the lighted verandah and the kids still up. Willing hands unloaded the car and carried in the bulging mailbag.

"Do you think Father Christmas will find our place?" queried little Jimmy anxiously.

"Of course," said Malley. "If he gets slewed, sure to be someone to give him a mud map."

Father Christmas never did let the kids down, and one year Trace noticed what she thought might be reindeer tracks among all the others down at the stock trough.

As each Christmas arrived, it marked a year of station improvements, changing staff and sometimes the inevitable accident which required evacuation by the Flying Doctor plane. In 1967 Bob's friend, Gary, came to live with us until they had both finished primary school, as his parents were leaving Alice Springs to manage the Wauchope Hotel and his mother, my friend Ada, would not have time to supervise his lessons. The two lads were handy young stockmen and great favourites with the Aboriginal ringers. Like most bush boys, at ten they could drive the vehicles, and when no men were available, they were often called upon to do a man's job.

A consignment of young bulls was trucked to the station and Joe gave the boys the job of taking them to the Wild Potato bore and tailing them out for a few days. I said I would drive their swags and tuckerbox to the Wild Potato yards and catch them up on dinner camp with a cut lunch. I got held up with unexpected visitors and it was five o'clock before I got to the yards. The bulls had already been watered and yarded.

"Bet you're starving," I apologised.

"No," they grinned. "We had lunch. We caught a snake on dinner camp, and it had eggs in it, so we lit a fire and had snake and eggs, and then conkleberries for dessert."

Joe was full of praise for them when they saved the day when he, Long Johnny and the two boys were taking the sale mob, due to be trucked next day, in to water at Wild Potato. The sale cattle had to be kept separate from the bush cattle also watering there. They were a couple of miles from the bore when Joe's horse threw him and galloped off. Long Johnny immediately set off after the horse but didn't get far before his own mount emulated Joe's and threw him too. There was no way the boys could hold the thirsty mob by themselves, so they carried on to the bore, held off the bush cattle while the sale mob watered and then took them to the yards a mile away. Only then did they go back to see how the men had fared.

"What if Dad had been hurt?" I asked, knowing full well that, in the same situation, I would have let the cattle go and raced back to Joe.

"I saw him on his feet, Mum. Johnny too. You've got to get your priorities right, you know!" said Bob. He was sounding more like his father every day.

The greatest call upon them came when we had the worst bushfires we had ever experienced. A thousand square miles of country were burnt out, and the first astronauts on their way to the moon reported the huge blaze they saw in central Australia.

It began when an Aboriginal ringer threw a firestick into a clump of spinifex to hunt out a goanna which had taken refuge there. He'd only been a week in the camp and had already shown himself out of step with the others. Malley and the men tried to contain the fire but it got away. When Malley reported it on the transceiver, Joe jumped into the new utility and hastily drove forty miles to the scene with fire-fighting equipment. They worked non-stop with the water tank on the Landcruiser and the next afternoon Joe drove the utility back across burnt-out country to try and assess the next move. Suddenly the fumes from the breather hole on top of the petrol tank caught fire and the next minute the vehicle went up. Joe managed to throw himself clear and crawl away before collapsing on the blackened earth.

It was nightfall before the ringers found him beside the burnt-out shell of the ute. At first they thought he was dead, but managed to revive him and brought him back to the homestead, by which time he'd recovered his sang froid. My first knowledge of events was when he walked onto the verandah with Harry behind him, saying, "He's blacker than me now, Missus, ain't he!"

He was too, from the scorched earth on which he had lain. Although his shirt had been burnt off and his eyebrows and the front of his hair were singed, he was otherwise unharmed.

After a quick shower and a meal, they were off again to take advantage of the cooler night to burn back across the advancing front. For four days and nights the team worked until they were absolutely exhausted and the Landcruiser was almost a wreck, but the fire was at last contained.

They all slept through twelve hours and then Joe sent them all to town to get the vehicle repaired and have a break. They left in the morning after transferring the water tank to a trailer. That night a wild electrical storm brewed up and as we sat at tea on the verandah we watched the vicious lightning strikes anxiously. Sure enough, the lightning started a fire we could clearly see, far too close to the homestead for comfort.

By nine the next morning the smoke began rolling in. We had no vehicle so Joe hitched the water trailer behind the tractor and ordered Bob and Gary to put on long-sleeved shirts and jeans and showed them how to operate the two fire hoses. All he could do was work along the perimeter fence of the homestead area to put out the burning grass before the flames got into the trees around the homestead. Tracey and Jimmy sprayed the garden hoses on the low bushes past the lawn. We couldn't see whether they were alight or not because of the thick smoke. I dashed inside and reported to the Alice Springs Fire Service and then relayed a quick call to Billiluna, but Gerry and the men were away, though Nola said she expected Gerry to call her on the transceiver at lunch-time.

It was a long, long day. Periodically the tractor came back to the stock-tank to refill the water tank and I dashed down to meet them with the tea thermos and sandwiches. The Flying Doctor Base operator told me to stand by on the transceiver and then told me a plane was coming with fire-fighters. Two hours later I heard the plane circling the airstrip but I knew the strip was well alight and there was no chance they could land. The plane flew away. If the worst comes to the worst, I thought, we can grab the pets and all jump into the stock tank. That thought was rapidly followed by another, that the surrounding heat would slowly boil the water and turn us into a horrid stew. The little ones turned on all the garden sprinklers and crept inside with eyes streaming from the smoke.

At one stage, Joe told me later, he had driven the tractor over a stump he could not see for the smoke and it was wedged there with the flames leaping around him. The boys turned their hoses on him and the tractor while he worked to free it, hoping that the water would last out.

At last, darkness fell and I could see the main danger to the house was over. Joe and two exhausted and filthy little boys came home, showered and changed and ate vast quantities of steak.

"Lie down on your beds in your clothes," Joe told them. "You can have an hour's sleep and then we'll have to go again." The boys were so tired they could only nod as they staggered off.

I poured Joe another cup of tea.

"If this was happening in a story," I said, "it would rain now."

He looked up, his ears cocked.

"It is!" he said.

I think the life-saving cloud sat over the homestead. We got twenty points. Enough, Joe said, to let the kids sleep till morning.

Gerry and the Billiluna boys arrived in the early hours of the morning; Malley and our team were back at piccaninny daylight. The plane came back in the afternoon and Joe went up with the fire advisers and overflew the area to plan his strategies. For the next ten days I baked bread and cooked steak to feed the fire-fighters at all odd hours. They worked mainly at night because the searing daytime heat made the job far too dangerous. In the long belt of mulga between the homestead and Wild Potato the flames leapt to the treetops and ran along the top in a matter of minutes, sucking up everything below in a roaring inferno.

Finally it was all over; there was nothing much left to burn. We hadn't lost any stock because the property was completely unfenced and the cattle were able to move freely ahead of the fire and back onto the already burnt country, but the hawks and the eagles had a picnic scooping up the small creatures fleeing from the face of the fires. The savage nightly storms continued but now the lightning bounced harmlessly off the bare ridges and the rain squalls teemed down. It only seemed days before the land which had been bare and black was covered with a mantle of green as far as the eye could see. Nature was making amends.

In school I was reading a chapter of *The Adventures of Tom Sawyer* to the children each afternoon. I ended one day with Tom and Becky lost in the cave with Injun Joe hot on their heels.

"I wish," said Gary mournfully, "that *we* could have an adventure!"

"Wouldn't you call your day on the fire-trailer last week an adventure?" I asked curiously.

"Oh, that!!" he dismissed the thought scornfully.

A few weeks later the last round of mustering and branding for the year got under way. Because the stock-camp was away from the homestead for six weeks at a time and could be as far as fifty miles away when they were on the far end of the run, Long Johnny was the obvious choice to leave at the homestead to offside for me and do the heavier chores, cranking the lighting plant engine and so on.

With Lorna gone and only myself and the children at home, I could rely on Long Johnny to entertain them after school and I didn't have to worry about adventurous kids getting themselves lost in the bush, as I might otherwise have done. Bobby and Gary were good company for Johnny too, sometimes accompanying him on long bush walks when their Saturday or Sunday chores were done.

One afternoon Brother Michael called in the Mission truck to drop off our mailbag which he'd picked up in town for us. What I didn't know was that he'd also dropped off two Aborigines at Johnny's camp. The boys told me at tea-time about the two men, one of them Long Johnny's younger brother, who were going to ask Joe for jobs in the stock-camp.

I wasn't unduly worried because there was a chance that Joe might be willing to employ them but I was a bit annoyed that Brother Michael hadn't seen fit to ask me if it was okay to leave them. I would have said no if I'd been asked. They lived on the Mission, after all, where Long Johnny's parents and other brothers and sisters also lived, and if Joe really wanted extra men he could easily slip over to Balgo himself to pick them up.

After I had read the children a bedtime story and kissed them goodnight, I settled down to read and answer the personal letters in the mail. It was about ten o'clock when I suddenly became aware of an uproar from Johnny's camp, violent shouting and abuse that almost drowned the noise of the lighting plant engine.

I realised instantly what had happened. I'd seen and heard enough of how O.P. rum affected men, white and black alike, in the bush. First it made them argumentative, then violently aggressive and then, if they drank enough of it, it sent them beserk. The blacks seemed to reach the berserk state quicker than the whites, which was the reason for the murders of so many Aboriginal women since the right to drink alcohol had become law. I recognised the aggressive stage of a rum binge and nearly wet myself with fright. I knew there was no chance of appealing to Long Johnny; his voice was yelling just as abusively as the others in a language I couldn't understand.

There were no locks on the three doors to the enclosed verandah. There was every chance that if they ran out of grog they might reason that Joe might have more in the house and come looking for it. Or worse! I propped a four-gallon drum against each door which would alert me by being knocked over if the door was opened.

Then I got the rifle from the office and loaded it. If Brother Michael had turned up then his life wouldn't have been worth a cracker. I reckoned that he had dropped off a couple of trouble-makers who'd played up on him in town.

I sat there with the gun across my knees for an hour, increasingly frightened as the bursts of profanity continued.

Then I got mad. If the engine wasn't put off soon — my usual job — it would run out of fuel and the fuel line would have to be bled before it would start again, a particularly arduous job with that engine. It suddenly became important to me to turn off the engine.

The homestead, the engine shed and Johnny's camp made a rough triangle with each side about 150 yards long. I took the torch and marched down to the engine shed and switched off the engine. The sudden silence was deafening. Then a muttering began. I yelled at the top of my voice, "If you bastards don't shut up and go to bed I'll put a bullet in your camp!" Brave words, because I didn't have the gun with me.

There was a torrent of words and then the sound of running feet, which stopped suddenly. That walk back to the house was one of the worst times of my life. They could see where I was by the torch beam; they could easily cut me off from the house and my precious children. Every instinct screamed at me to switch off the torch and run, but if I did they would know just how afraid I was. Adrenalin pumping, I concentrated on walking normally until I reached the door and slammed it behind me.

There was now only dead silence and that was almost worse than the fighting. At least I knew where they were then. Now they could be anywhere, right outside the door for all I knew. I waited another long hour before I went to bed, taking the gun with me.

At six o'clock the next morning, on my regular transceiver schedule with Joe, I told him briefly what had happened. "Don't do anything more, Honey," he said quickly. "I'll be right home!"

Two hours later I was relating the full story to him and being hugged and petted. I never saw him so wild as he was then. He strode down to Long Johnny's camp where the three were still sleeping heavily and kicked them awake. In minutes their swags were rolled and they were climbing sullenly on the back of the Landcruiser. Joe didn't tell me exactly what he said to Brother Michael when he got to Balgo. "I put the record straight," he said abruptly. "You won't have any more trouble from that quarter!"

I never saw Long Johnny again. A year later he rolled a utility while drunk and was seriously injured. I was very sad to hear it because, sober, Johnny was one of the best.

Chapter 15

The French Connection. Death at Chiller Well.

Although he always carried a sketch book with him, the only time Joe could do any serious painting was after the wet season set in and the road was closed for weeks at a time. The regular bore runs and the pumping during the wind droughts were no longer necessary with all the surface water about, and the cattle work could not begin before March at the earliest. It was the only time we had a quiet family life to ourselves, with Joe home most of the time and the employees away on holidays.

There was always the drama in early February of getting the older children back to school when the wet season was at its height and the roads and creeks were often impassable. At the beginning of 1969 I took all the children in a new car for a Queensland holiday, at the end of which we left Bob at All Souls School, in Charters Towers, to begin his secondary schooling.

Cyclonic rains chased us all the way back to Alice Springs, where Joe met us. He had been offered a ride to town in a charter plane and decided to take the opportunity to do some business and see Kim once again before she caught her plane to Perth for her final year at school. The rains which were to end a long drought in the Alice Springs district caught up with us there and it was five weeks before we could leave town. We only got as far as Chiller Well, where we stayed for a further two weeks.

Before leaving town, Joe sent a telegram to Bruce, the lone stockman holding the fort at home, asking if he wanted anything from town. "A long-haired mate!" was Bruce's cryptic answer. I complied. I employed a French lass, Jacqueline, and she was with us on our enforced stay at Chiller. The new owners, Sam and Beth Griffiths, also caught in town, had asked us to look in, if we got that far, to check on things for them.

We were in and out of bogs for three days and we knew we weren't going to get any further in the car when at last the little homestead came in sight. The white ants had made a fair job of consuming the one and only door, so Joe took it off and made a new one. He'd bought himself a new supply of oil paints and brushes in town but no canvas so he used the remnant of the old door and passed the time painting a picture of the homestead, which Beth still treasures today.

I had a term's school lessons in the mailbag, so it was lessons as usual on the Chiller Well kitchen table. We kept in daily touch with Bruce, until he finally decided he could risk a rescue attempt in the station Landcruiser. It was an eventful but successful trip home. It was a real wet year that: it was May before we could get the car back from Chiller.

Bruce was well satisfied with his "long-haired mate". He courted Jackie assiduously and a year later they were married in Perth, with plans to come back and start a roadhouse at Rabbit Flat, midway between The Granites and Tanami. Incidentally, the Rabbit part of the name refers, not to the bunny variety, but to rabbit-eared bandicoots or bilbies, which are fairly prolific in the area. Jackie had told me she was just passing through Australia on a world trip but she still lives at Rabbit Flat, probably Australia's most isolated roadhouse, where the traveller can sample true French cuisine.

That French cuisine was a bone of contention between Jackie and the children while she was with us. She showed them her recipe book one day. They had already learnt the French word for horse from her and here it was in her book. Cheval rôti. What did rôti mean? When she translated "horse roast", they were aghast. They loved their horses equally with their siblings and better than most other humans. Like a fool she admitted that she had eaten horse and told them there were special butcher shops in France that sold only horsemeat. Thereafter, they referred to her homeland as "that cannibal country" and often gave her a hard time.

It was perhaps paradoxical that, while we lived in such a remote area and were completely dependent upon ourselves for all the amenities of life which others take for granted, it was anything but a lonely existence. Our very isolation brought us to the notice of such a variety of research-oriented people that we spent many an evening during the long dry seasons in exciting and earnest discussions on subjects a world away from our own existence, as well as those in which Joe was the expert by virtue of his closer knowledge and observation of weather patterns, plant adaptations and the life-cycles of the indigenous animals and birds. Apart from the scientific researchers, we had numerous requests from overseas film units to make documentaries but knocked these back because they would mean too much interruption to our work and some of their proposals were patronising to the extreme.

We did, however, agree in late 1967 to the visit of the *Life* magazine reporter and photographer working on a special double issue called "The Wild World". They were the easiest of house guests and George Silk, the photographer, stayed for a fortnight, unobtrusively photographing most aspects of our lives. He handled our prickly seven-year-old Tracey like a veteran when, for the first couple of days, she stubbornly refused to be photographed. "Okay," he laughed, "I'll just take the others, then."

She held out for some days but then took to criticising him for "wasting all that film, click, click, click!", then to asking questions about his cameras and finally to participating whole-heartedly. His ploy was so successful that today Trace is herself an international journalist/photographer.

Unlike so many media write-ups, the *Life* article was spot-on and stimulated quite a number of interested letters from overseas. I have previously praised the diligence of the Australian Post Office and when a letter fell out of the mailbag, addressed to "Joe Mahood, Cattle Station, Australia". I thought, "Well, you couldn't do much better than that".

Fate smiled kindly on us for those years we lived at Mongrel Downs. Wet seasons came regularly and Joe's expertise in the cattle game was such that a reputation was soon established and store buyers at the Adelaide market sometimes told the agents that they would wait to buy until our consignments came down, when they could be sure the cattle would be quiet and well-handled, unlikely to fret and not eat as some of the wilder Territory cattle were prone to do. We used to sell fats too, cattle in prime condition to go straight to the butchers for slaughter, but the biggest percentage of our turn-off was forward stores, not yet prime, which were sold at Adelaide auction sales to buyers who put them on lush pastures and later resold them as fats. (Our last store sale in 1971 was a record one, with the highest average price ever paid for stores in South Australia and Joe received congratulatory telegrams from all the major stock firms in Adelaide.)

At the end of 1969 Kim won a Commonwealth Scholarship which would finance her way through university but, with our acquiescence, decided to postpone her acceptance and stay home for all of 1970. At that time she was interested in studying anthropology and the practical experience of working alongside our fully-tribalised employees could help to provide a good balance to purely academic theory.

Bill Wilson had bought Chiller Well from the Griffiths and Malley and his family had left us to manage Chiller, so Kim became an invaluable offsider to Joe in Malley's place. They both spent a great deal of time in the stock-camp and Kim, with Harry's help, set out to learn the Walmanjari language spoken by our ringers. Harry's wife, Daisy, was my house help and their baby, Anthony, was everyone's pet.

Kim, an expert rider and capable ringer, was a great favourite with the stock-camp Aborigines because they could rely on her to pull her weight with the best of them. They were a cheerful lot, who liked to extract as much fun as possible from the job and were always chiacking each other, Kim included. Harry's cousin, Mick, from Billiluna, a tubby young fellow, had now joined our team and Mick wasn't as game as his fellows, so he was often teased good-naturedly. Joe told me how, when he picked Mick up at Billiluna, his mother had anxiously pleaded "You lookim after my little boy proper, eh, Boss!". "Mick weighs about half as much again as any of the others so he ought to be able to look after himself," Joe laughed, "but I patted the dear old lady on the shoulder and told her I would. He's a good ringer and they all like him even if he is a bit of a mummy's boy."

Joe wasn't in the camp, though Kim was, when they were mustering near Lake Alec. Everyone else was quite used to the occasional sight of the lone old buffalo among the cattle but Mick had never seen or heard of him before. He was riding about a quarter of a mile away from the others when he was suddenly confronted by the old buff, which merely raised its head and stared at him. Mick gave a yell easily audible to all the others and raced flat-out for one of the few trees on the plain. It wasn't a very big tree but Mick went up it like a possum. It began to bend with his weight, which made him yell all the louder, particularly as the buffalo, intrigued by the noise, now began to lumber towards him.

The others rode over, laughing heartily, and Harry caught Mick's horse, but no way would Mick come down from his precarious perch until the buffalo had been shepherded away. It was much bigger than the cattle, a strange blue-grey in colour and with such an immense horn-span, Mick didn't know what he had struck and wasn't about to wait around to find out. It was quite a logical excuse but the others chiacked him unmercifully for days.

The following week Malley joined our camp temporarily to help draft sale cattle. His parents, with his younger brothers and sisters, were staying a few days at Chiller on their way from Halls Creek to Alice Springs and were company for Malley's wife, Oriel, and little daughter, Sandy.

Joe, Kim and Malley were all in the camp, two hours drive away, when I picked up a transceiver call from Yuendumu late one afternoon. "Nine Sierra Mike Lima. Nine Sierra Mike Lima. Urgent message for Nine Sierra Mike Lima!" Luckily, Joe had also tuned in on his portable set as he began to prepare the evening meal. The caller quickly told me that Malley's father, driving to Alice Springs with his family, had had a fatal heart-attack while at the wheel of their Landcruiser. He had collapsed against his wife, Betty, sitting beside him, but, miraculously, 13-year-old Lennie had managed to climb across from the back and reach into the cabin and bring the vehicle to a halt. They were then north of Yuendumu and Lennie had driven the vehicle there and now, all in shock, they needed Malley desperately.

I did not need to relay to Joe; he had heard the whole exchange and thought quickly. "Everyone is still out with the cattle. I'll get Malley and Kim and we'll be home as quick as we can make it. Have a meal ready for us. Betty will want Oriel with her, so Kim can look after Chiller while Malley's away. I can't leave here for long with the transports just about due for the sale cattle, or I'd go myself. See you in a couple of hours!"

I called Oriel, who had also heard the news, got the meal ready and packed some clean clothes for Kim. They got home and stayed only long enough to eat. Poor Malley was very upset. All he wanted to do was get to his mother to comfort her. Betty was half-Aboriginal and among complete strangers; she would not know where to turn. It would take him at least six hours to reach Yuendumu, after calling in at Chiller to pick up his family and drop Kim off. Malley and Kim left hastily in the Chiller Well vehicle, which had been left at the homestead while Malley was with the stock-camp, and Joe returned to Lake Alec and the cattle.

I had a six o'clock schedule with Kim every morning for the week she was at Chiller Well. It was a lonely place and she was just seventeen and a stranger to the Aboriginal staff there, but Malley had briefed her well during the drive there. She reported that a couple of the older men had tried her out the first day or so but, as she knew Malley's work standards well, they soon realised that she wasn't a new chum they could bluff. Nevertheless, it was a relief to me to hear her voice on time each morning and to know that she was not having any problems.

In emergencies such as this, neighbours pulled out all stops to help each other but on lesser occasions a spirit of competitive one-upmanship existed that made for many an amusing incident. Kim was returning alone from Alice Springs one summer day with a full load on the back of the Landcruiser when she came across some scattered sheets of galvanised iron spread across the sandy red-dirt road about twenty miles north of Chiller Well.

In the outback things literally did "fall off the back of a truck" at times and if they belonged to a neighbour you picked them up and advised their owner. If they were bound for a government Aboriginal Settlement, where we all knew stupendous wastages occurred, the bushmen reasoned that, as taxpayers, they had contributed to the purchase of the goods and, as consumers, they could put the items to much more productive use, so finders were keepers.

Kim couldn't fit the iron on the load; it was too heavy and bulky. Chances were that Malley might come along any time and claim the booty if she left it. The alternative was to hide it well off the road and come back for it later with an empty vehicle.

Because it was such a flat treeless terrain of low shrubs and bushes she had to drive quite a distance in the heavy sand, dragging the loot behind her. Then she planned to obliterate her tracks close to the road because eagle-eyed Malley would certainly investigate and follow up the evidence if he came along and saw the deviation.

So far, so good. She got the iron planted out of sight and then jumped back into the vehicle. The vehicle started but would not move forward. It wasn't bogged. After a few more fruitless attempts to go forward, she realised that it was time for the afternoon galah session, so she rigged the portable transceiver and called Joe. Knowing that Malley would probably be listening in too, she didn't want to say where she was broken down. If she did, obviously Malley would volunteer to come to her aid, being so much closer than Joe. She evaded Joe's questions about her whereabouts and concentrated on finding out what the vehicle trouble was. A few checks and tests and reporting back and the trouble was revealed. The four-wheel-drive gear was jammed in neutral.

She soon got going, wiped out the give-away tracks and reached home by nine o'clock, when she gleefully explained all to us.

"You'd better go back tomorrow and take Harry to help you load your spoil," laughed Joe. "Are you sure you can find the place again yourself if you wiped out the tracks well enough to fool Malley? It's about the most featureless stretch of road I know of."

"Of course I can," grinned Kim. "I checked the exact mileage between there and Rannehan's Well. I'm not a mug, Dad!"

Chapter 16

A Partnership Dissolved.
We Seek Pastures New.

I confidently expected to spend the rest of my life in the place which had become home to me in a way that no other place had ever been. I lived for the day only. But Joe was not quite so self-satisfied.

Our partnership agreement stipulated that, when the time came, we could buy Bill and Lorna out and pay them off over five years. Of course, the only way we could do this was to borrow the capital. Few partnerships survive indefinitely and ours was no exception. By 1971 it was evident that the time had come to part. A few years earlier Bill had had to split with Elders and persuaded us to agree to change our agents to Bennet and Fishers. We approached the Adelaide manager, who visited us with the local manager and agreed to carry us and pay the Wilsons out over five years. The valuation was made and for a whole week we lived on a cloud. A place all our own at long last! Our joy was short-lived. All unsuspecting, I copied down the dictated Flying Doctor Service telegram which advised us that a meeting of the firm's Board of Directors had taken place at which it had been decided to withdraw their offer of financial support. Our euphoria changed to immediate gloom. We knew the Wilsons wanted the place just as much as we did and had assets we didn't have to back up their request for a loan to buy us out.

For a few days Joe was pretty quiet and thoughtful. Then he walked over from the shed for smoko one day carrying a copy of *The North Queensland Register* with an advertisement he had ringed in pencil.

"How'd you like to live in Queensland, kids?" he asked.

"But, Dad," stammered Jimmy tearfully, "Mongrel Downs is the best place in the world!"

Tracey eyed him sternly. "Where the Mahoods are is the best place in the world, Jimmy," she corrected him.

"Yes, Dad, where is it?"

The advertisement was for a number of blocks to be balloted in a month's time, but, unlike the ballots of our early years, there were also three blocks to be auctioned. If we missed out in the ballot we could have another shot at the auction.

Bill chartered a plane and came out to offer us the alternative of selling our share to them.

"Okay," said Joe, "but you don't get the five years payout. I want $10,000 by Friday and the rest within a fortnight. Walk-in, walk-out in a month's time."

Bill didn't think he could arrange things that quickly but finally agreed when he saw Joe was adamant. The $10,000 was in our account by Friday and on Saturday Joe was on a plane to Queensland to inspect the blocks in Brigalow Area Three near the Isaacs River in Central Queensland.

At home — home for not much longer — I really felt the rug had been pulled our from under me. It was November. Hurriedly, I finished the year's schoolwork and faced the miserable job of packing our personal effects. Three bores broke down one after another but Harry, promoted at the beginning of that year to head-stockman, worked day and night to fix them. The lighting plant played up and I surprised myself by getting it going again. It took my mind off my disappointment for a while.

I thought Fate had played us a dirty trick and I wallowed in self-pity but little did I know then that exactly the opposite was the case. The Beef Depression was just around the corner and even our more affluent ex-partners were eventually obliged to sell Mongrel Downs. We would have had even less chance of survival with a big debt, high running costs and no income for years.

The stock-camp was still away mustering and branding. Bill sent up a man to stand by while Joe was away. We got a lightning strike and a fire on the horse-paddock fence about seven miles from the homestead one night but Bill's man, with Trace and Jimmy on the hoses, managed to put it out. When the new man elected to go to Chiller Well in the station vehicle to pick up some gear I sent Jimmy with him. He met some mates there and the grog flowed too freely and he passed out on the back of the vehicle about ten that night. Nine-year-old Jimmy knew I wanted the vehicle home in case of more lightning strikes and he could see the lightning playing on the north-western horizon, so he climbed in the cab and drove the 120 miles home, through the Lake Fidler sandhills at midnight, and got home at piccaninny daylight. His passenger awoke, vastly surprised to find he wasn't still at Chiller.

Trace was busy picking up and yarding our pet horses as they came in to the trough each evening to water. Her own particular pet, Li'l Pony, which she had raised from a poddy, did not put in an appearance and it looked as if we might have to leave without her.

Two days before Joe was due home, and four days before we were to leave, one of the Arab stallions brought his harem in to water and Li'l Pony was among them. How Trace cut her out I'll never know, but she slammed the yard gate seconds ahead of the rearing, roaring stallion and strolled back to the house well satisfied.

In the meantime Joe had looked over the blocks with an agent, decided which ones he'd put in for, collected a small branch of the brigalow tree we'd never heard of before, and flown to Adelaide, where he bought a Toyota Landcruiser from our old friend, Bob Sprod. He spent half a day at RM William's store, buying saddlery, camping gear and everything we'd need to start from scratch again and then he headed north.

Back at home he showed us the brigalow and the maps of the blocks. "What's spewey soil?" I queried wonderingly, pointing to the frequent annotations. We were to find out with a vengeance in the months to come!

Joe welded a framework for a canopy on the back of the Landcruiser while I packed the last of nine tea-chests of books. The rain set in just after Frank Fidler arrived to load up our gear and the nine horses, so we didn't waste any time. No last minute farewells to favourite places; no last view of Lake Ruth. Just as well. Neither did we say an actual goodbye to our ringers; they were still away mustering. Harry's wife, Daisy, with little son, Anthony, waved from their cottage verandah. Harry, who had been with us right from Joe's first year on the property, had earlier told us they'd be going to Halls Creek as soon as the muster was finished.

It was a rough trip. We met Bill at Chiller, on his way to take over. We detoured round a long boggy stretch on Mt. Doreen, didn't stop for lunch or tea, and arrived in teeming rain in the early hours of the next morning at Yambah Station, where John and Nancy Gorey had offered to spell the horses and store our gear until we were ready to leave.

In the morning the Goreys were ecstatic; the big home dam was brimming over. They helped us unload and we continued on into town to our house recently vacated by the tenants. The month we spent there was like living in limbo, not knowing for sure what our next move would be, and Spotsy and Paddy, our two dogs, who'd never known the restrictions of town life before, didn't like it any better than we did.

We met Kim and Bob, home for their university and school holidays, and tried to explain to them all that had happened. I hadn't wanted to tell them before their final exams because I knew that Bob, particularly, would find the news hard to take.

The ballot was drawn in early December and once again we had missed out. Joe flew to Mackay for the auction a week before Christmas. He would only consider the one block with a bore on it and our financial limit was very little over the upset price. We waited with bated breath for his telegram.

Here rumour worked in our favour. In Mackay the word went around that Joe was a rich grazier from the Territory and his determined bidding gave the impression that the sky was the limit.

The last opposing bidder fell out just before our limit was reached and the block was knocked down to us for $54,000. The other comparable block went for twice that. Joe paid three-tenths of the purchase price, with ten years to pay off the rest, and for the improvements, the bore and some ancient fencing. He flew home and we whooped and danced round the house together.

"The country was settled a hundred years ago," Joe told us, "and I think all the improvements were done in the first ten. It's going to be a challenge!"

"Did you find out what spewey soil is?" I asked.

"No, I forgot. I guess we'll know soon enough." He was right there!

Joe had given eight years of his life to establishing Mongrel Downs and the children and I had been with him for all except the first year. It was home to us as no other place had ever been because the challenges and the responsibilities were so much greater. Now we had another virgin block and a different set of challenges to face. Joe was adamant that we must never look back, never make comparisons, only think ahead. So we turned our backs on the past, firmly banished our regrets and thought with confidence of the future.

Chapter 17

Queensland Greets us
with Cyclones and Floods.

The Territory had been our home for more than twenty years but it was surprising how quickly we could finalise our affairs there. Parting from close friends was the hardest thing to do. We transferred our bank account to Mackay and asked the house-agent to rent our town house for us until he could sell it. Harry had asked Joe if he and his family could come to Queensland with us and Joe had agreed to contact him and pay the fares from Halls Creek. I phoned the Aboriginal Affairs official in Halls Creek only to discover that Harry was in jail for a month and no offers to pay fines were acceptable. The best that we could do was to ask that Harry be informed there'd be a job waiting for him when he got out; he'd have the initiative to find out how to contact us.

It upset me to think that the man who'd been such a terrific backstop to me while Joe had been away — mild-mannered, capable and more responsible by far than some white employees we had had — should be languishing in jail like a common criminal. It was a fair guess that, because he'd gone to Halls Creek with a good cheque, he'd become involved, despite himself, in an Aboriginal booze-up and the almost inevitable brawl. With Daisy's previous record before she'd ever met Harry it was likely she'd played a big part in it. I tried to find out where she and Anthony were but nobody seemed to know.

We left Alice Springs on the 28th of December 1971. The previous afternoon we had loaded our gear onto Frank Fidler's semi-trailer and as an afterthought Joe and Bob had pulled down the garage in the backyard and loaded that too. Cyclone Althea had devastated Townsville on Christmas Eve and Joe thought it could be some time before we could get the planned machinery shed with living quarters onto the block and erected; the garage would do as shelter in the meantime.

We loaded the nine horses and the rest of our gear at Yambah, where the Goreys gave us lunch and refused to take payment for agisting the horses. Then it was all aboard for Queensland. Kim travelled with Frank in the truck. The Toyota, now with a tarpaulin canopy on the back, was packed with the swags over the suitcases to make comfortable travelling for the passengers and dogs. We stopped for a quick meal at sundown on the roadside near Barrow Creek. At Wauchope Joe and I stayed an hour with friends to let the truck get well ahead and then we drove all night, each taking a turn at the wheel while the other stretched out on the back among assorted kids and dogs. We had to think of the horses and do the trip as quickly as possible and the arrangement was that we wouldn't stop until we caught up with the truck. That wasn't until daylight and I was never so pleased to see anything as that truck pulled off the road and Frank and Kim with a fire going and breakfast ready. Frank had driven all night but he looked as fresh as a daisy. The meal and the short spell soon brightened the rest of us.

By ten o'clock we had crossed the border into Queensland. We got a permit to travel the horses from the Camooweal policeman, and by early afternoon we were unloading the tired horses for a feed, a drink and an overnight spell at the Cloncurry saleyards. Same for us humans and I added a shower and a change of clothes for myself and the two younger children at the CWA Hostel. I rinsed out our dirty clothes and took them back to our makeshift camp at the saleyards and spread them on bushes to dry. We bought fish and chips for tea and boiled the billy and it wasn't long before we all fell into our swags. It had been a long day and we wanted an early start in the morning if we were to make Charters Towers by the next night.

We didn't get as good a start as we planned as no way was Bob's horse, Blackie, going to get back on that truck again. The other horses filed on obediently but we lost a good hour persuading Blackie and had to put up with the comments and advice from two old fellows who sat on the top rail and sneered at our efforts.

The dirt road had only been re-opened the previous day. It was pretty chewed-up after the rain but we had a good run and a welcome stop at Gilliatt Ponds for lunch. The shady trees were welcome too, after the treeless plains we'd traversed all morning. The horses were parked in the shade and Paddy the sheepdog meticulously scratched away the pebbles at the foot of a tree to make a comfortable resting place while Spotsy stood stiff-legged, watching him. Paddy was just making the circular turns the way dogs do before settling themselves when Spotsy stalked over and growled softly. Paddy abjectly conceded his place and Spotsy settled himself sneeringly in the vacated nest. This was a bit much for Joe, who had been watching Paddy's careful preparations and never could stand bullying. "You mongrel thing!" he shouted, putting the toe of his boot into Spotsy's backside and ejecting him smartly. Paddy rather reluctantly resumed his place at Joe's order; I think he knew he was going to have to pay later. Paddy was no hero; the easy way out was always good enough for him and he'd kow-tow to anyone so long as he could keep his hide intact.

Near sundown we reached bitumen road again, flanked by emerald green grass as far as the eye could see, the result of the cyclone rains of only a few days before. It was one o'clock in the morning before we finally reached Charters Towers, found the saleyards, unloaded the horses in the moonlight and made camp.

Next morning was busy. We bought a spare drum of fuel, replenished supplies and enquired about the state of the road south to Clermont. The estimates were about 50/50 as to whether we could get through or not. After lunch at the pub (our last meal under a roof for many months) we reloaded the horses, this time without any trouble, and set off. Our spirits were high. The country looked marvellous, the cattle we saw were fat and shiny and when we got to the Belyando River the bridge showed just clear of the swirling water.

We stopped on the northern bank at a newly established roadhouse and while Frank, Joe and I had a cup of tea the kids went down to look at the swollen river. They came back dripping wet, saying non-committally that they'd had a swim. Well, it was a hot day and bush kids often went swimming in shorts and shirts, so it wasn't until we were crossing the bridge and I saw the raging torrent that my heart went cold. Impetuous Tracey, all her childhood spent in the land of occasional lakes but no rivers, had jumped straight in, luckily close to the bank. Kim and Bob had dived in simultaneously; one caught hold of a bush and then the hand of the other. As Tracey surfaced, they grabbed hold of her and gradually manoeuvred all three to the bank via the bushes, while Jimmy watched breathlessly. I could never prise any real details out of them, but when I next crossed that bridge in the dry time I saw that their "bushes" were actually the tops of tall gum trees.

There were no tracks on the dirt road ahead; ours were the first vehicles to cross the river after the rain. We drove on without mishap until sundown, snatched a quick meal, and reached our turnoff just north of Clermont well after dark. We camped late that night and left the horses on the truck.

New Year's Day 1972 dawned clear and after three hours' driving we came to the turnoff to the Fitzroy Development Road, breakfasted there and then tackled the last leg of our lengthy journey. Our destination was Bombandy homestead, where the owner, Alistair Brown, would direct us to the track which led to our block.

After almost two hours' travel on the heavy track, we turned left at a set of wheeltracks through thick timber. The Toyota negotiated the stretch of heavy black soil and the small creek spilling over from a nearby dam but the truck bogged. Not just an ordinary run-of-the-mill bog, we saw with sinking hearts, but one which threatened fair to swallow the truck before our eyes. It was only ten o'clock in the morning but every hour we worked with long-handled shovels and dragged-up logs, only saw the truck sink lower and lower.

At mid-afternoon Joe and I went the rest of the way to the homestead in the Landcruiser where we met Alistair and Mrs. Brown and their long-time stationhand, Jim Somerville. The two men came back to the bog with us. Obviously we weren't going to get the truck out that day so we jumped the long-suffering horses off and Kim and Bob saddled up and, with a mud map drawn by Jim Somerville, set off to follow fences and creeks to take the horses through Bombandy country until they reached the fence which enclosed the only paddock on our block. The rest of us followed in the Landcruiser; it wasn't a trip I'd like to repeat.

Straight out of the desert, it took a bit of getting used to the thick, grey green forest like a wall in front of us, no horizon to see and the narrow track disappearing periodically into threatening pools of dark water. Then at last a wire gate in the sagging fence and we passed through just as the sun came out for a few brief moments and turned the grass to emerald and brightened the surface of a small dam where the horses were already drinking. We stood on our own, our very own, land at last and if the immediate future did not offer the latest in modern conveniences the long-term prospects were all that we'd dreamed of.

Bubbling with enthusiasm, we battled our way back to the bogged truck and a late meal. Most of our gear was unloaded and covered with a tarp, which was just as well because we'd barely eaten when it began to rain again. That night we slept where we could in the cabins of the truck and landcruiser or burrowed under the edge of the tarp, but a violent thunderstorm and the crashing of lightning-struck trees jerked us all awake about 2 a.m. When the rain tapered off to a light drizzle Frank got a fire going with a tin of petrol tossed onto some gathered branches and made coffee. The sandflies and Scotch Grey mosquitoes zoomed in. We'd never seen mosquitoes like that before. Joe said he thought they were small bats at first, only no bat had a bite that savage.

Then the rain began to belt down again and the mosquitoes went wherever mosquitoes go to escape death by drowning. We fled to the vehicles' cabins again. Our radio was packed somewhere in the mound of gear, so we hadn't yet heard that Cyclone Bronwyn was stationary just off the coast and that all the rivers and creeks in the district were once again rising in angry flood.

By late afternoon the following day the truck was finally extricated from the mud, the gear reloaded and the rest of the hazardous journey to the homestead achieved with only one minor half-hour bog on the way. We were all mud-streaked, sweat-stained, mosquito-bitten and generally filthy from our exertions. The most desirable thing in the whole world was a good shower first and then a meal. We all took it for granted that what was mandatory hospitality in the Northern Territory would also apply in Queensland. Even feuding enemies got a clean-up and a feed at each others' homesteads back in the NT and a place in the shed or on the verandah to throw out the swag if the weather was inclement. After Mr. Brown had shown us where we could park the vehicles and departed without further comment, it took us a while to realise that they did things differently this side of the border. By then it was dark. Joe lit a campfire and I scrabbled around in suitcases to find clean clothes for everyone. By torchlight we found the creek and performed our ablutions as best we could. Trouble was that once the caked mud was removed the mossies and sandflies had a bigger area of skin they could get at.

An hour later we all sat clean around our campfire at our belated meal, leaning into the smoke that made our eyes sting but deterred the mossies somewhat. It was the only time I ever saw Frank cranky, and that was on my behalf. He kept referring to the hundreds and hundreds of weary travellers and complete strangers to whom I'd given bathroom access and meals at Mongrel Downs, so much so that in the end he made us all laugh. Though it rained again that night we all slept soundly, dog-tired but full of enthusiasm for the morrow.

Despite the sheeting rain that greeted day two with us just about but still not quite on our own land, we decided unanimously to make the final run for it — but in the Landcruiser only. There wasn't the least doubt that the semi had come as far as the terrain would allow, rain or no rain, and it was definitely going to be a ferry job for our gear for the last stretch.

Swags, tuckerbox, suitcases and, by lucky chance, the roof purlins of the little garage made up the first load. We had a mud map of the track to the bore where we planned to establish camp. Nothing but horses and a short-wheel-base Landrover had ever negotiated that track before and we could already see that there were going to be some very tight pinches for the Landcruiser between the trees. Desertwise, we all knew the need for a permanent camp close to water but right at that moment making it *out* of the water seemed a more important priority to me and the next few days didn't change my opinion much.

Ten miles, seven bogs and eight hours later we reached the bore site. The men took the tarp-covered frame off the back of the vehicle, tied it on to the purlins and parked the suitcases and swags on the purlins out of the mud. The tuckerbox and a bag of sugar were shoved in after them and when the next downpour began the two little kids, the two dogs and I crept in too. There wasn't much room to spare. The others decided to go back for a second load as the boggy places were now all corduroyed with logs. Before they left, Joe got a fire going underneath a huge half-hollow fallen log by syphoning a fruit tin full of petrol from the fuel tank and pouring it over some wet kindling under the log, followed by a lighted spill of paper wrapped round a stone and tossed from a safe distance.

The rain roared down on our shelter and the thunder blasted away, but the intermittent lightning bolts lit up the gloom enough for me to see that the water was creeping in underneath the tarp and rising steadily against the purlins. I manoeuvred the suitcases up onto the swags but it wasn't until I felt my feet suspiciously sticky inside my boots that I remembered the bag of sugar. The kids and I were sitting on the edges of our swags up to the ankles in gradually-thickening syrup.

Night fell, the rain slackened off to a drizzle and we crept out and put the billy on our very satisfactory protected campfire. Two hours later four bedraggled figures staggered in out of the darkness; the vehicle was bogged a quarter of a mile back and they'd had to feel their way through the dark guided only by the smell of smoke from the campfire. We did full justice to the meal and gallons of hot tea as nobody had eaten since sun-up.

Luckily we were on open ground near the bore. We didn't plan on throwing out our swags too near any of the big trees whose branches crashed down at intervals with the more savage wind gusts. We laid out the swags in strategic places and then shared the shovel around until everyone had cut channels to carry the water run-off around instead of over us as it flooded down the slope to the creek beyond the bore. With our swag flaps folded over, we slept comparatively dry, comparatively being the key word.

The next day the sun sneaked through the cloud cover now and then and for a week the rain was limited to one late afternoon downpour each day, timed just right to wash off the dirt and sweat of the day's labour from clothes and bodies.

By the end of our second day at Cattle Camp, we had built the large bough shed that was to be home for the next twelve months, made a reasonable bush fireplace and dug the Toyota out of the bog. Included in that load was a table, which made meal preparations a bit more civilised - no more enamel plates floating away, no more horses stepping in the butter. I was lyrical with gratitude for the forethought in bringing that table on the early load. Joe said that if I kept the tucker standard up I might even get some chairs by the end of the week. Another bonus we all noticed was the absence of flies. Back in the desert it had always been purgatory to try and eat out of doors any time after sun-up; this was, as Frank said, like a picnic every day of the week.

It was a wonderful time of discovery. Some of the trees, grasses and plants were familiar to us but many more were not. There were insects and birds quite new to our experience and a welter of varicoloured fungi and the kangaroos were grey instead of red. The goannas weren't as big as perentis but they were fatter; the pythons were more in evidence; flying foxes bickered in the trees and the wild pigs frightened me stiff. To balance the absence of bush flies, there were ticks, fleas and leeches, but the bird and butterfly life was magic.

Each day more loads were ferried to the growing stack beside the bough shed and at the end of a week a successful effort was made to get Frank's truck back to the main road, although it took a full day to do it. Kim left with Frank to go as far as Townsville to enrol for University there and then make her way back to Mackay as best she could.

The rain came in again and it was another week and many bogs later before the last of our gear and the shed components were ferried in from Bombandy. Joe and Bob erected the little shed quickly and stowed all the tea-chests inside. I knew nine of them contained our lifetime collection of books but by now I was rather vague about the contents of all the rest.

Kim sent a message to Bombandy that she was now back in Mackay and, as our food supplies were running low, Joe and Bob decided on a quick trip to Mackay. In three hours there Joe bought a motor-bike, chainsaw, stores and a pressure light and ordered a small bulldozer and a workshop shed with living quarters to be delivered after the Wet and when we'd found a suitable track to get a truck right to the site. In the early hours of the next morning the elated travellers got home, having miraculously forded the last creek in driving rain and slippery mud.

That night we had a round-campfire conference to decide on the name for our new home and many were the fanciful suggestions, which Joe finally halted by announcing conclusively "We're camped on Cattle Camp Gully; we'll call the place Cattle Camp!". I noted it in the diary, along with the definition of spewey ground — an area which appears dry and hard, unrecognisable from the surrounding terrain until you drive onto it, break through the crust, and find yourself trapped in a bottomless quicksand of almost liquid mud.

Fourteen years later, in 1964, Joe was still a desert nomad, but on his own property, Mongrel Downs.

...For almost a year, Joe had no permanent headquarters, throwing out his swag each night and building his campfire wherever the job of the moment took him.

Joe Mahood, bull shooting, 1949-50.

[1950] . . . was the first year payment for Aboriginal workers was introduced, a pound a week for stockmen and ten shillings a week for house-girls but, as they now had to buy their own clothes, blankets and tobacco, nothing had really changed much for them.

. . . Joe was later stunned to find out that . . . none of the administrative staff of the Native Affairs Department, who made and applied the policies, seemed to have the remotest idea of the social system or the tribal laws of those over whom they played God.

Aboriginal Stockmen - Mongrel Downs. Painting by Kim Mahood.

. . . the sandstorms occurred with dreary regularity. One blew for three days and, when it was over, the five wire fence around our yard showed only two wires above the sand.

A typical Centralian sand storm, Finke 1958.

There are times, perhaps once in a hundred years, when the deserts of the inland are suddenly flooded with enormous lakes, up to 20 ft deep. In 1966 the bare claypan of our airstrip was transformed into Lake Ruth.

. . . a small, white-sailed yacht scudded along with two ten year-old boys in yachting caps at the helm, and two more swimming behind. On the white beach three teen age girls in bikinis sunbaked and smaller children were building sand-castles.

The Exclusive Tanami Yacht Club, 1966.

... *Albert Namatjira was living at the settlement at this time and was already quite famous as an artist. What a genuinely likeable gentleman he was! ... Joe was now quite dedicated to a start in oils so he gave all his watercolour gear to Albert, who was profuse in his thanks. A couple of days later he offered, as Joe wouldn't take any payment, to paint a picture for Joe instead.*

Joe Mahood, Self Portrait.

Joe and Marie - The First Shelter on Cattle Camp, in 1972.

... *Our way of life was something you couldn't put a price on; it was the sweet icing on the prosaic damper of hard bush toil.*

Joe and Marie outside the shed which was destined
to be their home for 13 years. 1974.

At an earlier Christmas at Cattle Camp, 15 years before.
... *the boys drove home at sundown with a big, fat goanna. Tracey stuffed the goanna with rice and herbs and baked it in the outside oven. We had a 'traditional' dinner at midday.*

Christmas Day at Cattle Camp, 1988.
Carolyn, Jim, Bob, Kim, Tracey and Joe.

Chapter 18

The First Year on Cattle Camp.

The first year on Cattle Camp was, all things considered, one of the best years of our lives. We lived in a big bough shed with the small garage erected beside it to protect our personal gear; we cooked on an open campfire; we worked from piccaninny daylight till sundown. Yet none of these seemed like hardships because we were working for ourselves on our own land. Even the toughest challenge seemed just a matter of Joe, with his analytical and inventive mind, figuring out what we had to do and the team swinging into action. As always, we appreciated the privilege of living close to nature and the land and as we gradually adapted to the change from desert dwellers to forest dwellers, there were daily rewards of observation of the forest wildlife.

At the end of January I took Kim to the James Cook University in Townsville and delivered Bob and Tracey to school in Charters Towers. I came home with a second-hand kero fridge, a prized addition to the bough-shed kitchen-dining area. In the far corner stood Jimmy's school desk which we eventually walled off with leafy branches to prevent the too-frequent visits of one, two or more horses. The dogs were allowed; they kept the snakes away.

Joe's special brand of humour maximised our highs and minimised our lows. As he slumped, well-nigh exhausted, by the campfire one night I said, "Where's the bloke who claimed he could lick his weight in wild-cats to win me for a bride?". He gave me a rueful, lop-sided grin as he flexed an aching muscle. "He's taking a spell, Hon," he said, "Right now, he'd be flat out chopping down a mushroom! Ask me again tomorrow."

On the credit side we had the camp for food and shelter. We had the bore, two small silted-up dams and a few semi-permanent waterholes in Devlin and Scrubby creeks for water; and there was a 14 square-mile paddock to hold the nucleus of our cattle herd. There were no debits, only challenges. Once the old rosewood posts and snaggled barbed-wire fences were repaired or replaced we were in business. The original old wagon road from Clermont to Mackay had passed through that paddock. Hardly a sign of the old road now but there was an old wooden gate near the Devlin which we named Welcome Gate as we set about trying to entice inside some of the wild cleanskin cattle which roamed the surrounding forest.

A cleanskin a day
Keeps the creditors away!

chanted the kids on their return to the camp after a successful foray.

Their scores began to add up, but still Mum hadn't contributed. Then, returning from town one day with a loaded vehicle, I spotted a black cow with a big weaner and a little calf lurking along the fence a hundred yards or so from the gate. With Tracey planted behind a tree near the opened gate I went bush in the vehicle. A hectic few minutes over logs, between trees, dodging anthills, and I got my quarry through Welcome Gate. That night I could skite round the campfire too.

We fenced a holding paddock near the bore and in April we bought our first cattle: 12 Braford herd bulls at $500 each, 90 cows with weaner calves at $150 each and 68 heifers. We cross-branded them lazy ⋛Z6, the 6 for number in the family, the M for Mahood and the Z for as far as we could go.

During March Joe had made a track through the scrub wide enough to take a truck to bring in the small JD 350 bulldozer he had bought. This also meant we could now get the new shed materials to the homestead site we had chosen up on a tongue of red tableland country about a mile from the bore. We also bought two semi-loads of scrap steel from a sugar mill near Mackay. This scrap-heap was Joe's pride and joy; there was no end to useful equipment and handy inventions he manufactured from it over the years.

But first things first. Fences, yards and loading ramp had to be built and, unlike the Territory yards, these yards had to have a dip. The dozer took off the top layers of earth but the rest of the dip hole was a crowbar job through rotten-sandstone, rotten both by definition and human reaction. Joe welded the steel dip sheets together and the finished job looked like the prow of a ship as the dozer towed it towards and into the hole and the assembled family cheered mightily.

While the kids were home for the May holidays, we got the framework up and the roof on the new shed. Then Bob dropped his bombshell. He asked us to let him leave school on the grounds that he would never again have the opportunity to learn all that went into establishing a station from virgin country and obviously his father could use a strong young offsider. He was fifteen, had topped Grade 10 the previous year in English, Science and Manual Arts and we had always taken it for granted that all the children would complete secondary school and probably go on to University. We were finally persuaded when he agreed to complete his secondary schooling by correspondence and sit for the Senior Exams. (For the next three years he was to study each evening after a hard day's physical work, at first in the bough shed with a pressure lamp and its attendant myriad insects and later in our homestead shed. I was very proud of his dedication when he passed in all subjects and gained his Senior Certificate.)

When we borrowed $33,000 from the Commonwealth Development Bank and bought another 139 cows and calves and 13 cows from Lancewood Park, the time had come to think of more paddocks, so we went looking for the surveyed corner peg and blazed trees which marked our northern boundary.

Joe parked the Landcruiser in a likely small clearing and pulled out the map marked with compass bearings. Some of the blazed trees had succumbed to past bushfires but after some hours of hunting through the almost-featureless scrub we found the line, but the corner peg continued to elude us. "It ought to be right about here!" said Joe, puzzling over his compass. It was. Right there underneath the vehicle, hidden by the long grass.

A Mackay air-charter pilot delivered their mail to Bombandy every Tuesday so we arranged for ours to be sent there too and on one weekly visit Joe chartered the plane to take a run over the whole block ($15 for three-quarters of an hour) to get some idea of where to put fences and future watering points. The freehold title of the land was conditional on boundary fencing, two additional waters and clearing most of the brigalow scrub.

In July we organised to get the first thousand acres cleared and in September bought another 22 cows and calves and 51 cows from Logan Downs and 125 head of cows, calves and heifers from Wyoming and Cairo. Among that last draft was "the Cairo cow". No wonder they sold her. Fences meant nothing to her. She had mastered the technique of crawling under any fence ever designed. She just lay down and wriggled under and she taught every calf she had to do the same.

Joe observed her teaching her weaner one day. She went under and back three times before the calf got the hang of it and went under too.

There had been no rain at all since March and the local people were now talking drought, but we were not yet unduly worried.

A telegram came from Harry, now in Alice Springs and keen to join us. The agent there still hadn't sold our house and we needed the money, so we decided that Bob and I should drive over, sell the house and pick up Harry and his family. Our domestic water supply was contained in two forty-four gallon drums, filled at the bore and transported to the camp on the vehicle. We topped up the drums for Joe and Jimmy, advised them to "go easy on the bath water" in the absence of the vehicle and set off, co-driving on the long stretches.

Bob had been driving on properties for years but, being only fifteen, had never driven on a bitumen road before. He was highly impressed at the ease of driving after his dirt-road-only experience but we had a dicey moment when he was at the wheel between Cloncurry and Mt. Isa and a policeman on a motorbike pulled us over. He only gave Bob a cursory glance as he warned him to pull off the road to let a wide load with a house on board go past.

We were away ten days in all. We sold the house, picked up Harry, Daisy and little Anthony and I bought Bob a smart pair of tan stockman-cut trousers, (which I eventually paid for three times, when he sold them to Tracey and I had to advance her the money and again when she sold them to Jimmy under the same circumstances).

When we arrived back at the camp in the early hours of the morning Joe and Jimmy put the billy on for the weary travellers. They still had a quarter of a drum of water left - mission accomplished with a day to spare.

It was great to have Harry and Daisy with us again, and Harry's bushcraft in tracking cattle soon came into its own, despite the terrain being different from his desert experience. Jimmy greeted four-year-old Anthony like a long lost brother.

Now we began to understand what the locals meant by drought. In the Territory we had been used to many months and even years between rainstorms but not to the very different effect it had on forest country and east-coast cattle watering, not at bores, but at dams and waterholes. The cows, some heavy in calf, began to bog in the receding mud. We bought licks and molasses as the locals advised us to do but nine cows, too weak to recover after they had been pulled from the mud, died over a period of a couple of weeks and I acquired my first two poddy calves.

Brian Moohin brought the molasses in a tanker and pumped it into a small tank near the yards and a number of forty-four gallon drums nearby. Then we repaired to the new shed to unload other gear, only to return to find that a cow had turned on the tank tap with her tongue and a wing of cattle were paddling around in the spreading pool of molasses, lapping like puppy dogs. That was one of the few times I ever saw Joe lose his cool; he reckoned Bob hadn't turned the tap off properly and called him "a useless bloody jackeroo!". He had to apologise a couple of days later when he discovered that the cows had unscrewed the caps on the drums, which he had himself tightened, and lowered the contents a good tongue's-length.

Early November storms brought some respite but also lightning-strike bushfires and we lost a few more weak cows. Now there was a new sense of urgency to our work, racing to muster and brand calves, to complete the horse paddock and Scrubby Creek fences, to finish the new shed so we could move in, and to organise the aerial seeding of the cleared scrub. Whenever anyone reported seeing cleanskins beyond the fence limits the boys downed tools and rode after them, usually successfully, so that we more than replaced the cattle we had lost in the boggy dam.

The girls came home on holidays and joined the workforce and I ferried home some basic second-hand furniture, a new deep freezer and a second-hand lighting plant. I was especially pleased with the little old wooden dunny I'd also acquired when a neighbour went modern, — a great leap forward from the drum and plank set-up behind the tarp at the old camp. We left ten-year-old Jimmy to dig the requisite hole for it, while we all went fencing one day.

When we returned after dark there was only a silent building and no welcoming light on to greet us. A pathetic "Help! Help!" came floating across the flat. We should not have laughed at the over-enthusiastic little boy who had dug the hole too deep to be able to climb out, but we could not help ourselves.

At last the shed, or what the manufacturers called the workshop with temporary living-quarters — two bedrooms and a bathroom up one end — was completed and a wood stove in place not far from the back door. Joe claimed that I wore a pad a foot deep walking round that stove and patting it whenever I visited the site to check on the building progress. The small shed we had brought over from Alice was put on skids and dragged up to the site for a private home for Harry and Daisy and Harry added a neat verandah.

We moved in and the two poddies which had slept each night either side of the campfire moved with us. The freezer meant that we could now get a killer and keep the meat fresh, so Joe shot a beast drover-style from up in a tree and we all gorged that night on rib-bones. When the meat was all packed away in the freezer and Joe's favourite brisket had been salted, he made a rope from the hide. Nothing was wasted on Cattle Camp.

When the wet season rains began in earnest, we were all under a roof again. We ate together at our place; I cooked and Daisy washed up. We had a great Christmas, with a tree the boys brought in on Christmas Eve and the decorations from the box I'd brought from Mongrel Downs. Expectations were high. Little calves were dropping everywhere, grass growing inches overnight, and we hardly gave a thought to the repercussions on rural Australia of the December election of the first Federal Labor Government for 23 years. That was all to come. Joe played "Silent Night", "Six White Boomers" and other favourites on his guitar and we toasted old friends and new friends in home-made ginger beer.

Chapter 19

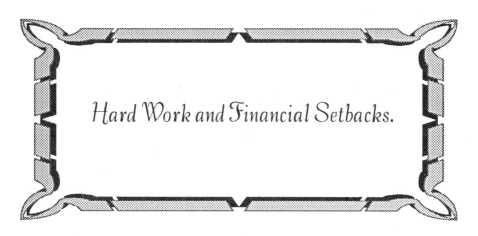

Hard Work and Financial Setbacks.

Our second year at Cattle Camp began with rain and ended with rain. The season was kind to us. We had a reasonably comfortable home and a cattle herd and the stock agents, Elders-G.M., were once again willing to back us. Our main project for the year was fencing. We needed to fence boundaries with neighbours and subdivide paddocks in order to manage the cattle effectively. Also, our very basic yards needed to be extended and a race and loading ramp built.

The only fly in our ointment at that stage was the new Federal Labor Government. Labor governments had never been exactly sympathetic to rural landholders and this one wasn't set to be any different. The new Treasurer had already taken to referring to us in Parliament as "fat cats" and "hillbillies". I thought that was a contradiction in terms though, in our case, I conceded that there was a hillbilly element in our way of life when we heated a boiler of water on the open-air stove and carted it inside to a tub and drew lots for first bath.

The first setback was to Kim's education plans. She had won a Commonwealth Scholarship to university which had covered all her expenses but now, with a year still to go, the new government had cancelled the scholarships and there was no way, with our previous year's tax assessment based on our sale of half of Mongrel Downs, that she would be eligible for the new plan of assistance to all cash-strapped students, regardless of scholastic merit. We were now well in debt and couldn't help her, so in the end she had no option but to find a job and try to complete her degree part-time.

Now, with the withdrawal of full tax deductions for capital expenditure and improvements we had to completely re-assess our financial situation. The hike in interest rates from 6.5% soon to 12.5% didn't help with our debts to the Development Bank and Elders-G.M. and to top it off we got a tax assessment of over $10,000 to pay from our ex-partners' accountants.

This really stunned us, as we had sold our half of the cattle at book value but the Wilsons' accountant had assessed them at market value. What we lacked, the local Tax Department advisor explained to me, was a signed agreement from our partner before the sale that the cattle would be assessed at book value, and obviously we weren't going to get that in retrospect. He was very sympathetic but I was pretty close to hysterical as I thought of all Joe's non-stop hard physical work going down the drain to pay a tax bill that had been manoeuvred onto us because we were so naive and inexperienced in business dealings. "It doesn't happen often," the tax official told me, "but we have been known to waive tax under certain circumstances. I suggest you apply."

Our new accountant said he'd only ever heard of one successful appeal. He wasn't hopeful; he said it could drag on for years. Then out of the blue, I received a cheque for $5000, a legacy from an aunt's estate. I only had it between the two pick-up and delivery mail services; I re-addressed it straight away to the Tax Department for my share of the bill. We applied for a hearing for Joe's half, and some months later we were interviewed and Joe's payment was waived.

During January, with the girls both home on holidays to boost the workforce, we added to the yards to the point where we could muster the cattle and cross-brand the rest of our bought cows and brand the weaners and new calves.

Branding was a team effort. Harry lassoed the cleanskins and big weaners and tied the end of the rope to the Toyota bullbar. I backed the Toyota away to drag the beast up to the bronco-panel. With Harry on the front leg and Kim and Bob each grabbing a back leg Joe applied the brands and castrated the males. Tracey passed the brands and kept the fire hot and Jimmy did the ear-marking. The little calves were scruffed in the best Territory traditions.

Between regular rainstorms the musterers flushed out the cattle from brigalow scrub and forest hide-outs with many a wild ride and many a graphic post-mortem discussion on the wiles of cunning old cows bent on selling their freedom dearly. Joe opined that Brahman-cross cattle were a lot smarter than shorthorns and they could think like a horse and kick like a camel. When it was time for Tracey to go back to school, our tally was 362 cross-branded cows and 315 branded, mostly big weaners, and before the heavy rains began in earnest in mid-February we had added another 79 cows and 68 weaners.

They were busy days. Cooking for eight or nine hungry mouths was a job I shared whenever I could bribe or bludgeon anyone else to do it and every night after tea I held a school session to teach Harry and Daisy to read and write. Harry was one of the most rewarding students I ever had. We continued the evening lessons for the whole time he was with us and soon progressed to simple maths and geography and Australian history. In return he told me the Aboriginal legends his grandmother had told him as a little boy. Daisy's interests were more in the art line; she had quite a talent for painting birds and flowers.

Isolated by the heavy rains I began to worry that Daisy's expected baby might arrive while we were cut off from town. She wasn't at all perturbed. Having offsided for me at Mongrel Downs when I had been a midwife by necessity she said she had complete faith in me. I was far from sharing her opinion and besides, Kim had to get back to university, so when the rain let up for a couple of days, we made our epic trip through miles of mud to the Beef Road. Harry steered the Toyota and Joe dragged us through the worst patches with the little bulldozer. Twice it was touch-and-go whether the dozer would bog too and we were all well and truly mud-splattered travellers by the time we reached more negotiable terrain.

We booked in at a Mackay motel and the Hall family were delighted with their first experience of TV and hot water straight from the shower; Kim and I revelled in the luxury too. The next day we farewelled Kim at the railway station. I bought stores and a second-hand wardrobe and some chairs and Harry bought a cot for the baby and then we left Daisy at the CWA hostel and headed for home. We only got bogged once in a breakaway on Bombandy. It was good timing because Cyclones Kirsty, Leah and Madge were all brewing further north and in the Gulf. Baby Jennifer arrived three weeks later, a real little doll, already complete with a full head of beautiful black curls.

Meantime the work went ahead, cutting fence posts on the high country. Joe had found some good stands of rosewood in Top Paddock reasonably close to the Amaroo boundary to be fenced first. He also noticed sandalwood and silky oak trees among the rosewood on a further ridge. I minded Anthony while Daisy was away, supervised Jimmy's lessons and tuned in to the weather broadcasts. Once again the Isaacs, Connors and Funnel Creek were all in flood and the mosquitoes and sandflies were in plague proportions. The insect bites we all suffered affected Joe most of all and time after time became infected.

When he took a break to go to the first Nebo sale for the year, it was the first time he had left the property since the preceding July. He yarned with stock agents and neighbours and thoroughly enjoyed the contact with the outside world. On the agent's advice he arranged to sell some cull cows at the April sale in Emerald.

He wasn't long home before he discovered that the Three Day Sickness epidemic spreading through the district had reached our herd; it was apparently spread by sandflies. If left completely undisturbed, a sick beast might recover but if harassed at all it quickly died. We lost two or three good steers and did not muster again until the sickness seemed to have passed. Joe built the loading ramp at the yards so we could load the cattle onto trucks for the forthcoming sale.

Our stockhorses had worked hard for many months so Joe bought two more, Lucky and Gift, from Bombandy at $40 each, to help share the workload. The boys mustered about 700 to 800 head over two days and Joe drafted and culled and ordered the trucks for the Emerald sale. The next day he noticed a sick beast in the yard and on further inspection found seven or eight more — a second round of Three Day Sickness. He cancelled the trucks just as they were about to leave their base and the next day we listened on the radio to the excellent prices achieved at the sale. Oh well, next time!

In the meantime, a spell for the workers. The Shannon family, of Saltbush Park, was celebrating one hundred years on the property with a picnic race meeting and all the neighbours near and far were invited. Joe elected to stay home and catch up on bookwork in peace and quiet. The rest of us piled on to the Toyota in high spirits and headed through the scrub to Amaroo where we met the Huttley boys in their utility and they took the lead to the Bee Creek Crossing. There we met the Chaffeys from Seloh Nolem (spell it backwards) in their car. Ross, with one leg strapped from thigh to ankle, was on crutches. The Crossing was obviously a four-wheel-drive-only job. Car and utility were parked in the shade and everyone piled onto the Toyota for the last leg of the journey to Saltbush Park.

What a great day. Nearly all the races were calcuttas, horses supplied by Saltbush Park and auctioned to an "owner" for the race. The young fellows jockeyed and chiacked; the settlers swapped problems and solutions; the old hands advised and the wives admired Daisy's baby, swapped girl-talk and tried to keep track of excited kids.

Reluctantly we decided to give the evening dance a miss so we could tackle the return trip over the Crossing in daylight. Once across, we boiled the billy, broached tucker-boxes and had tea together in the growing dusk.

It was towards midnight when the Cattle Camp contingent finally arrived home. Joe had had an equally satisfying day. After a cursory glance at the station bookwork he had decided to delegate it to his number one secretary — me — and had turned to his easel and oil paints. For one whole day he had pushed all work plans and finance worries to the back of his mind and transferred some of his colour-note pencil sketches of trees and clouds onto a glowing canvas.

Communication with the outside world became an increasing problem. We now had a weekly mail delivery to a mailbox at the Bombandy turn-off but that was 22 long skinny miles away and in times of flood rains the mail run did not operate. There was a party-line telephone at Bombandy and in an emergency the Browns would take a message and send their stockman to deliver it, but this was an imposition we did not like to continue.

The previous government had allocated money to extend the telephone service to all the Fitzroy Development Scheme settlers but this was cancelled by the new government. In expectation of the service we had not outlayed the money on a transceiver but now it was obvious we should have to do so. We ordered a portable model and Joe erected the tall mast that had characterised all the outback stations of our Territory experience. Our Royal Flying Doctor base was Cairns. Now we could send and receive telegrams, contact a doctor in an emergency, chat with neighbours who had also bought transceivers, and Jimmy could join in the School of the Air lessons.

The August Federal Budget had brought us no joy. The money we spent on improvements of fencing, clearing and water supplies had previously been fully tax-deductible in the same year but now only a small percentage could be claimed. Fuel went up 7 cents a gallon, subsidies for country air services and rural newspaper postage disappeared, a one cent per pound charge was imposed on export beef and producers had to pay a levy for the Brucellosis and TB campaign.

This unexpected financial setback heralded the beginning of one of those cyclical downturns familiar to all who live on the land — a time to pull in the belt and mark time. For the established properties this might be an option, but for the battlers on virgin country with all their improvements still to do, the prospect was devastating. As yet, no one even had a real homestead; we were all still living in caravans, sheds or tents.

"Okay," said Joe. "We re-assess our priorities. We make do with the fences we've got and the house goes from year four to future. We can't afford to employ extra help now, so the jobs will just take a bit longer to do."

Luckily, at that stage, we had no hint that worse was yet to come with the beef market crash of the following year. (That wasn't a belt-tightener; that was no belt at all and the pants right down round the ankles!)

As 1973 drew to a close, Joe made his first trip to Mackay for 18 months and spent the last of our loan money on 29 heifers, followed by the purchase at Charters Towers of 55 Braford heifers and 98 Brahman heifers, all at top prices. We planned to recoup the outlay with the sale of our first draft of fat young steers.

The waterholes in Scrubby Creek, rapidly turning into muddy death-traps for thirsty cattle, were fenced off and the cattle watered at either the Devlin waterholes or the trough at the bore. From the bore Joe laid a pipeline up to the shed to augment the supply in the rainwater tank and did the appropriate plumbing for sink, laundry tubs, bath and shower. He reckoned it was time for a few home luxuries seeing the temporary status of our living quarters was no longer definable. He added a bonus for Mum — a new garbage-disposal unit. He cut the top out of one of his empty 44-gallon drums ear-marked for troughing and set it up with the grass cleared away from around it so I didn't set the countryside alight when I burned it off.

Bob's exams for three Senior subjects were imminent; he had to go to Nebo school to be supervised by the teacher there. Inevitably, the storm clouds threatened. We were just tossing up as to whether we should send Harry with the Toyota to get Bob out before the road was closed if it rained heavily or whether to hang on another day, because if it didn't rain we'd need the vehicle to check the waters for the cattle.

"Vehicle coming!" announced Jimmy. It was the Nebo Elders agent, Alan Bignell, with whom Bob was to stay during the exams.

"Looks like rain," said Alan. "Thought I'd better come down and pick you up early in case you got stuck getting out tomorrow."

Privately, I was thinking that Joe ought to go too as he'd staked his knee chasing cleanskins in the brigalow a few days before and was limping painfully. However, I knew from long experience that the only way I'd get Joe to a doctor was if he was unconscious and couldn't argue, so I held my peace on that score and just thanked Alan for his forethought.

The heifers Joe had bought at Charters Towers arrived on trucks two weeks before Christmas. We drooled over the beautiful creatures and admired the smooth hides and long floppy ears of the pure Brahmans. Harry, Daisy and their children were to go with the truckies as far as Charters Towers to catch a bus there for Alice Springs for a holiday. It was a sad parting because we knew, deep down, that they weren't going to return. Harry insisted he'd be back but I knew that Daisy yearned for the companionship of her own people and resented Harry's easy ability to mix on equal terms with whites and his long-term plans to save money and educate his children.

Harry had a big cheque; Daisy couldn't see any further than splashing it on presents and grog for relations and friends, in line with the genuine socialist philosophy of sharing that was part of her Aboriginal heritage. I could see both sides and I knew Daisy was going to win once she was on her home ground. As it turned out, there was no way we could have afforded to employ Harry the following year, though he did write to us a couple of times saying that he hoped to come back "sometime".

Tracey was home for the holidays and my nephew, Steve, same age as Bob, had come from school to spend a fortnight with us before going on home to Darwin for Christmas. We cross-branded the heifers and the first little calf was born while they were still in the yards. Then the team mustered the steers we had arranged to truck for the last sale before Christmas.

It had long been my practice to lay in enough rations to see over the whole wet season if need be but this time I did not see the sense of making two trips when I had to go to Mackay on the 20th of December to put Steve on a bus for Darwin the next day and meet Kim's bus from Townsville. I could buy the stores and Christmas presents at the same time.

Cyclone Una changed all plans. The rain began on the 18th, Una crossed the coast the next day and the Wet began with a vengeance. (It was the following March before we could get a vehicle in or out of Cattle Camp, and then only just.) Joe let the steers go and cancelled the trucks.

Our world became a grey curtain of steadily pouring rain. I was glad we had the transceiver for outside contact and gladder still when I had to contact the Flying Doctor in Cairns because the cuts on Joe's hand had turned septic. The long-ordered Flying Doctor Medical Chest had not turned up, so the doctor said that Joe would have to get to a doctor in Mackay and arranged for a charter flight to Bombandy airstrip. This time Joe only put up a token resistance, the hand was so swollen and painful. We started out with me driving the Toyota but only got a mile or so before I bogged it hopelessly. Bob went for the bulldozer to drag me out and bogged that too. Joe didn't speak at all, just watched. When it was obvious that we weren't going to rescue the dozer that day he just put his head down and started to walk. We all trailed along behind him. I wondered whether the little plane would actually make it over the clouded ranges and if it did whether we would be able to cross the flooded Devlin to reach the airstrip. As far as I could see, it was either antibiotics for that hand or amputation.

When we came at last to the crossing, Joe just ploughed on through the water without a break in his step, as if he hardly realised it was there. We followed. One of the dogs got swept away but managed to hook his paws round a fallen tree trunk so Bob was able to rescue him. Steve towed the other before the water could snatch him too.

The plane was waiting on the airstrip. The anxious pilot stared into the lowering sky and hustled Joe aboard for the take-off. We shouted goodbye but he didn't seem to hear us. On the return trip to Cattle Camp I was so relieved that Joe was going to get to a doctor that I hardly noticed the knee-deep slog for miles through the clinging mud or the non-stop attacks of mosquitoes and sandflies.

We had only had the transceiver for three months but it was to be a lifesaver more than once. Within a month a young settler, caught in the floods, was not to be so lucky. None of his neighbours knew that he was a diabetic or even whether he was on his property at the time. When a neighbour finally did ride a horse up to his hut, he found that its owner had died alone at least a week before, unable to get the medication he needed or to make his predicament known.

With our transceiver and the radio we learned that the floods were widespread. Joe was in Mackay, Kim was marooned in Townsville, and this would be the first Christmas that the whole family would not celebrate together.

Tracey and I checked our meagre rations while the boys smeared themselves with home-made mosquito repellent and set about digging out the Toyota. I gave Bob the rifle with a request to bring us home a turkey for Christmas dinner.

We had a few pounds of flour but no store yeast, so I appropriated the lone last potato to make potato yeast. A tin of ham and a tin of peaches were reserved for Christmas dinner and Tracey combined brown sugar, cocoa and the dregs of various containers to make some quite delicious toffees and fudge for the boys' presents. She had made the Christmas cake at school and brought it home with her. She made a quick sortie into the bush and came back with a hatful of yams to bake in lieu of potatoes. Then she begged a cupful of yeast and made a batch of ginger beer and six bottles of raspberry soda coloured with cochineal. From that time on Tracey's status as the best cook in the family has never been challenged.

It took the boys all day to chop down trees and corduroy a track out of the bog and they drove home at sundown, bringing with them, not a turkey, but a big fat goanna. They also brought a Christmas tree, salvaged from those they had chopped down for the corduroy, which they erected in a bucket of sand at the end of the shed. Jimmy fell asleep while decorating it.

Tracey stuffed the goanna with rice and herbs and baked it in the outside oven during intermittent showers. We cut the cake for smoko on Christmas morning and enjoyed a traditional (except for the poultry) dinner at midday. Jimmy had found some bonbons left over from the previous year among the decorations so we even had paper caps. The two dogs sneaked inside and partook of their last tin of Pal, saved for the occasion. Zip the cat was most appreciative of the goanna scraps.

After tea that night we talked and played records. To save the batteries I put the radio on only for the news. It was all about the floods and the Middle East oil crisis. When we let down the nets over our beds and extinguished the lamp hundreds of fireflies winked on and off under the roof above us, a splendid display just like a miniature starry sky, which we hadn't even noticed while the light was on.

On Boxing Day Bob and Steve packed lunches in their saddlebags to try to ride the Scrubby fenceline but were home in the early afternoon to report that the horses were bogging and it was impossible to go further. A lot of country was right under water.

I tuned in for a telegram from Joe on the 27th to say that he had chartered a plane to Bombandy for the following day to bring him home and pick up Steve. He had booked Steve on a plane for Darwin on the 29th. The kids left early to go to the Bombandy airstrip and were able to drive nearly as far as the corduroyed stretch. Steve left his luggage for me to return when I took Tracey back to school, taking only a knapsack with a change of clothes. The others returned at lunchtime, reporting that they'd left Steve at the airstrip. At mid-afternoon when we finally heard the plane, they set out again to meet Joe, who had brought potatoes, onions and coffee, all he could carry comfortably. He also had the Flying Doctor Medical Chest but that had to be left at Bombandy until we could pick it up with a vehicle. Steve told us later that he had sat by the airstrip in full view of the Browns' verandah from 10am until 4pm but no one had acknowledged his presence. I had taken it for granted that they would have given him lunch or at least a cup of tea and was annoyed with myself for not thinking to pack him a lunch to take with him.

With the boss fit again, he and Bob got a killer (and a turkey), checked fences and floodgates and began work on a new wing for the yards. On New Year's Eve our rainfall tally for the year was just on 40 inches, 15 inches above average, and it was still raining heavily.

Chapter 20

The 1974 Floods.
The Collapse of the Cattle Market.

Our first priority for 1974 was an airstrip. There was a relatively open strip of firm red ground on high country about a mile from the shed. Joe and Bob marked out the length, cut down the taller trees at either end which might be dangerous for landing and take-off, cleared the anthills and bushes from the strip itself and flattened the tall grass by driving the vehicle up and down, towing a heavy triangular metal framework behind it.

It rained every day and Devlin Creek was soon too wide to consider any more trips across it to Bombandy. The last trip was on the 8th of January when the kids had heard the mail plane and made the crossing to pick up our belated Christmas mail bag and some items from the Medical Chest.

On the 21st we sent a telegram giving the coordinates of the airstrip to Pioneer Airways in Mackay and asking for a charter plane to bring us some more rations and to pick up three passengers. I was going to take Tracey back to school and enrol Jimmy as a boarder at All Souls, Bob's old school in Charters Towers.

When Charlie, the operator at the Cairns Flying Doctor Base, took my telegram, it was obvious that he was becoming increasingly perturbed at the reports of rainfall readings from the stations near the headwaters of the Gulf rivers. They were higher than those recorded in the big flood year of 1956 when the Gilbert River had run ninety miles wide. Charlie spent all day repeatedly calling the stations along the Gilbert and Einasleigh advising them of what was coming and trying to arrange evacuations. A helicopter working from Georgetown began the epic task. As daylight waned many upstream homesteads were already flooded, paddocked horses drowned and cattle trapped against hundreds of miles of fences. Heavy rain was forecast for the next two days. One station manager, on his homestead roof with a transceiver, reported a brown swirling sea of water into which the swimming heads of his star racehorses disappeared one by one.

By next day the floodwaters of the Gilbert and Einasleigh Rivers had joined together, 150 miles wide, and the operation to rescue marooned families from homestead rooftops and threatened out stations was in full swing.

Nothing of this rated even a mention on the State news radio and TV services, but when Brisbane itself was inundated a week later the entire news reports were devoted to the city's plight. Luckily the local radio news in the provincial cities kept us all in touch with what was happening in the north. The Brisbane flood, worst for a century though it was claimed, was only a trickle to what was happening in the Gulf. From the air a photograph showed an apparent coastline 200 miles inland from where it really was.

Meanwhile, we waited anxiously for the charter plane, which we knew could only leave Mackay if a break occurred in the heavy clouds which obscured the ranges behind the city. Only a fortnight before, one of the Pioneer pilots had crashed his plane into the mountainside but had miraculously walked away unhurt, although the plane was a complete write-off.

We were once again out of food, reduced to one weevilly packet of cake-mix I'd found that morning at the back of a cupboard. For lunch I sifted out the weevils, added water and baked 18 dubious-looking small cakes. We ate them very slowly, making each one last as long as we could. Soon there were three left on the plate and five famished faces staring at them.

"Very nice, Honey, but I've had enough," said Joe.

"Yes, me too," I said.

Instantly three hands flashed out and snatched and then three young faces looked somewhat abashed because the hands had betrayed just how hungry their owners were.

Then, as the last crumbs disappeared, we heard the drone of an aircraft. We raced up to the airstrip in high excitement as the little plane circled and made a perfect landing. "Can we slip across to Bombandy to pick up our Medical Chest?" asked Joe. The pilot agreed and Joe climbed aboard. I could read his mind. If there were going to be any more medical emergencies for him he planned to stay put and treat them himself with the antibiotics supplied in the Chest.

When I boarded with Tracey and Jim twenty minutes later I thought to myself that this drama of getting children back to school in the wet season equalled any of our Territory experiences. I sat beside the pilot and stared down in amazement at the sea of water below us studded with islands of higher country and isolated homesteads. The pilot seemed a taciturn young fellow, not much inclined to chat. As we approached the ranges, luckily still visible, I said,

"I suppose you know the pilot who had such a lucky escape here a couple of weeks back."

"Yep, that was me!" he remarked.

I quickly shut up and let him concentrate on his flying. The rain held off until we landed but sheeted down again and drenched us as we raced for the hangar.

It took us another two weeks to get to Charters Towers. We were stranded for two days on the coastal train. We had a few welcome days with Kim in Townsville and then went by rail motor for the last leg of the trip.

As I enrolled Jimmy at the school a helicopter landed in the grounds with two grubby little boys who'd been plucked off a homestead roof in the Gulf country. Others had arrived the same way but, as yet, only half the school's full enrolment had turned up.

Jimmy, who'd never been to a regular school in his life, was suddenly overcome by second thoughts.

"I'm too little to go away to school!" he told me pathetically.

I thought I was doing him a favour by asking for him to be put in Page House, where his cousin Steve was Head Boy. Steve would keep an eye on him. (Jimmy later claimed that that was the lousiest thing I ever did to him because Steve *did* keep an eye on him — in situations where anonymity would have far better suited Jimmy's ideas on how to make boarding school life tolerable for an eleven-year-old boy.)

I only found out years later why Jimmy's schoolteachers looked vaguely surprised and murmured about pushing a child beyond his capabilities when I chose university prerequisite subjects for him in Grade 9. As a new boy Jimmy handled his IQ test the same way Joe handled most of the government forms which regularly came his way. Joe had a theory that most government paperwork was a useless waste of time and paper, designed only to give government clerks something to do to justify their existence and that nobody ever read the completed forms anyway. To test his theory he filled in the most ludicrous, ridiculous and provocative answers he could think of — none of which were ever queried. The older children sometimes assisted in the exercise and Jimmy was only following a family tradition when he tackled his first official form. Of course, Jimmy's answers *were* read — which explains why his teachers didn't expect too much from him in academic achievement.

As I caught the rail motor back to Townsville I mused, as fond mothers do, on how my bright little boy would soon be earning his teachers' approbation, probably at the same time as the teachers were registering an IQ of about 20 and deciding not to push the poor little fellow "beyond his capabilities".

The steam-train journey back to Mackay was only held up by washouts twice. Now that we had an airstrip at home we could be included in the weekly mail run, so the February 5th flight found me cramming myself and all the cartons of food that would fit into the little four-seater plane. (Food supplies for Cattle Camp was to be a regular cargo for some time. What with Cyclones Wanda, Yvonne, Zoe and Alice on line until mid-March there wasn't a long enough period between rain depressions for the road to dry out enough to take the weight of a vehicle so we depended on the plane. I would telegraph a list of urgently required items to the pilot who did the shopping for me, and for other stations on the run, at no extra charge. Those bush pilots were worth their weight in gold to us, flying in all weathers and always so cheery and obliging. When the loss of the government subsidies finally put all the small bush airlines out of business they were sorely missed by the people they'd served for more than a generation.)

What a surprise when the plane landed. There was Malley Brown with a wide grin and arms wide to greet me and Joe laughing beside him at my pleasure. Back at the shed were his wife, Oriel, and their children, his mother, Betty, and her friend and Malley's young brother and sisters — eleven in all. Their two cars were bogged at the Seven Mile on Bombandy, the same place where we had come to grief on our first trip with Frank. Malley had walked in to Bombandy, got directions and walked to Cattle Camp. Joe went back with him to guide the party on a short cut with the men carrying the smaller children. It was after dark when they got back to the shed to a welcome meal and a sharing of swags and blankets until every-one was bedded down. It was like a delayed Christmas with such welcome visitors, who had tackled that muddy 15-mile hike through tall timber and long grass, tormented by mosquitoes and sandflies all the way.

Malley was on holiday from his job managing a Territory station, where he had left Bill McKell as caretaker, the very same ringer who had split up the crocodile of schoolchildren waiting for their X-rays at Finke so long ago.

For two days we reminisced and caught up on the fortunes and misfortunes of our common outback acquaintances. One morning I tuned in to a Territory transceiver channel on which we could sometimes receive but rarely transmit. McKell's familiar drawl floated through, loud and clear, reporting his location and predicament.

"Only the top of the cabin stickin' outa the mud," he said mournfully. "The lousy sheila's wrote me orf and I'm three 'undred long skinny miles from 'ome. Oh well, what Malley don't know won't muck up 'is strides as long as I c'n get back before 'e does!"

"You're fired, McKell! You're fired!" yelled Malley into the speaker. All transmission stopped dead.

Malley hastily packed up his family and Joe was able to drive them a good part of the way to the bogged cars, which were extricated by nightfall. They had half of North Australia to cross back to the station, and half of that was probably under water. Malley, no doubt, was agonising all the way about what might be happening at his deserted homestead.

There were no more attempts at road travel on Cattle Camp for some weeks.

In mid-March Kim came home on the plane. She had tried for over a year to carry on with her university course and support herself with a part-time job but the stress of missed lectures, not enough time to study and having to hitch-hike at all hours to the University because she could no longer afford to live on campus was finally too much. We were delighted to see her.

"I've got just the job for you," said Joe. "Seven days a week, start at six and knock off at sundown. No pay, of course, but the tucker's thrown in and the boss takes his turn cooking it!"

"Done!" said Kim.

She slipped easily into the vacancy left by Harry's departure and also made it possible for Bob to take more time off to study for the two new Senior subjects he had to complete by the end of the year. However, there was a problem, common to all the settlers in our district during the coming beef depression. Someone on the place had to get a job that *did* pay or the mob didn't eat.

"Who can we spare?" said Joe. "Who's the most useless on a horse or in the yards?"

Naturally they all looked at me. I glared.

"Rephrase that!" grinned Joe. "Who's the least able in the stationhand department?" I still glared. He chuckled.

"Rephrase again. Who's the most able in this family to earn a decent wage at an off-the-property job?"

"Fair enough," I said, "I'll go!"

It was easy to get a teaching job, thanks to the recent, but very short-sighted, demands of the Teachers' Union that young appointees no longer be obliged to sign a bond to work for the Department for a period equal to the time they had been paid as trainees. (It was short-sighted because, from that time, the trainees were no longer financed by the Education Department, nor was there any obligation to employ them when they graduated.) As a result of the bond removal many of the new breed of teachers worked only until they had saved the fare to Europe or Bali, and a bit over, and then quit. The State Government was recruiting teachers from the United Kingdom, Canada and USA in a frantic effort to fill the gaps.

I was soon appointed to teach high school French and English at the newly established coal town of Moranbah, accommodation provided, starting after the May holidays. It was close enough for me to get home for the weekends — we'd worry about the means later.

The necessity of an off-property income was heightened by the rapidly deteriorating cattle prices. By late March the value of a beast had dropped by 40% on the prices of the previous December. In normal years the Nebo sales began in March but this year the first sale was repeatedly cancelled at the last minute due to rain and each week prices fell in general. When the sale was again postponed from the May 11th date, Joe told the agent in frustration that we'd sell our steers on the place and take what we could get. The job of repeated mustering, drafting and dipping for a sale and then having to let the steers go again was wasting time better devoted to post-cutting, fencing and laying a poly pipeline from the bore to the turkey-nest tank Joe had built with the dozer about five miles away in Big Scrubby paddock.

The first buyer who Alan Bignell brought to inspect the steers wanted a higher Brahman content but the second buyer on the 29th of May agreed to take 84 head at $102 each and produced his branding iron to cross-brand them in the yard. The trucks to take them away arrived three days later, having successfully negotiated the track which Joe had patched and rebuilt over washouts and bog by many hours work with the dozer.

The first truck was duly loaded and pulled away from the ramp and parked to wait for the next truck to load. "Hang on!" yelled Kim, pausing in her efforts to push reluctant steers up the loading race. All eyes followed her gaze. The loaded truck, with the added weight of the cattle, was slowly and inexorably sinking before our eyes through the hard surface crust into the wet earth below.

Joe, Kim and Bob got their horses and jumped the cattle off. Joe got the dozer and pulled the truck out of the bog. The truckies had smoko with us and said they'd try again in a fortnight if it didn't rain in the meantime.

The cross-branded steers were remustered and trucked on the 14th of June, exactly six months later than we'd first planned to sell them. The price we got was less than half our first estimate but $20 a head more than the current saleyard price when they were finally trucked away. As the last truck left Joe leapt in the air and clicked his heels and said he thought they'd be worth that extra $20 to the buyer because, with all the mustering and yarding up they'd had, they'd have to be the best-handled mob of steers in all Queensland.

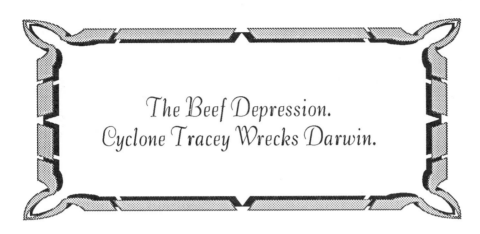

Chapter 21

The Beef Depression.
Cyclone Tracey Wrecks Darwin.

I was never any good at making bread. My efforts ranged from just tolerable to practically uneatable. I couldn't master the knack of kneading the dough properly. No one in the family would ever volunteer to make a batch because they all knew full well that, if they did, they would be elected unanimously as permanent bread cook. Instead, they just grinned at each effort and said, "Nicely cooked, Mum, nicely cooked!".

That didn't make me feel any better, either, because we all knew the story about the stock-camp ringers who found themselves without a camp cook and drew straws for the unpopular job, with the agreement that anyone who complained about the food had to take over the job himself. After a few days the reluctant loser began to serve some pretty awful meals but still nobody complained. Finally, in desperation, he made a stew of spuds, onions and the biggest cowpat he could find. The first ringer to take a mouthful spat in horror and yelped "Shit!" but quickly recovered himself to add, "But nicely cooked, mate, nicely cooked!".

Thus, my first reaction to the prospect of an off-the-property job was high glee that I wouldn't have to make bread any more. I could buy it in town and bring it home on the weekends together with any other supplies we needed.

This is not the story of my career as a schoolteacher but, suffice it to say, I enjoyed it greatly and, over the years, met some marvellous kids. We solved my transport problem easily. Every Friday afternoon Joe would drive to the junction of the Development road and the Peak Downs Highway and wait for me. I never had any trouble hitching a ride to the turn-off because both teachers and miners from Moranbah often went to Mackay for the weekend. I could usually arrange to meet my Friday driver at the same place for the Sunday afternoon return trip. I treasured my weekends and particularly the drives with Joe, when we swapped the separate experiences of our week apart.

The first weekend was a sad one. Spotsy, warrior dog, true childhood companion of Bob, Tracey and Jimmy, was no more. Since our early arrival we had periodically seen the huge tracks of a dingo, half as big again as normal and Jim Somerville, from Bombandy, had mentioned the giant dog's attacks on calves more than once. On the Tuesday evening a dingo bitch had lured Spotsy into the bush and set him up for the attack. Unlike dogs, dingoes mate for life but the dogs don't seem to know this and are always susceptible to the flirt of a pretty red tail in the undergrowth. Spotsy limped home the next midday, gashed and bleeding. Joe stitched the terrible wounds and gave him an antibiotic injection and he lay quietly on a blanket in the shed. Despite the awful loss of blood Joe thought he might recover.

About nine that night Bob noticed that Spotsy was missing. At first they thought he had only gone outside to relieve himself but a search failed to find him. When he returned next morning he could only crawl. Many of his stitches were split and he was bloodied from more savage slashes. Joe laid him on his blanket and bathed the wounds but he died two hours later. The warrior dog was buried under a big bloodwood tree near the shed. We never again saw the outsize dingo tracks; Spotsy had gone back to even the score.

Life settled into a pattern. Joe, Kim and Bob mustered, dipped and branded, laid miles of poly water pipe, and carted sand and gravel to concrete another two bays of the shed floor. Both Kim and Bob took casual jobs of a few days' mustering or a fortnight's droving whenever the opportunity occurred.

Tracey and Jimmy came home for the August holidays. At the beginning of the month the Labor government had revoked the rural fuel subsidy which had applied for many years. Inflation was rising out of sight, strikes were rife in many industries and there was no prospect of any overseas beef market. Luckily for us, the Remote Area Education Allowances were still intact (though a subsequent Labor Government would revoke them also). Together with the similar State Allowances and the free student train travel a good percentage of the boarding school costs were still covered.

The season continued to favour us and our social life improved when I invited some of the young teachers home for the weekend. Most of them were to become permanent friends. They all owned cars which solved my transport and saved Joe two trips to the Highway and they all, particularly the two English girls, had romantic notions about cattle stations, heightened by Joe's songs and bush yarns. They were very easy to entertain and helped us remember the attractive aspects of our lives to cancel out the decidedly unromantic realities which now and then cropped up.

Joe the Optimist decided to try for a further Development Bank loan and when Lee Curran, the bank representative, came for an inspection, we were introduced to that nasty word "viable" which, at some period of their lives, haunts the nightmares of most Australian farmers and graziers. It all boiled down to the fact that if the bank thought we were going to make it, we'd get a loan. Lee was impressed with Joe's improvements and the likely productivity of our land and he left us quite hopeful that we might qualify for the $12,000 loan we needed to pull and seed the brigalow country as part of the prerequisite for the freehold title.

In good heart Joe began the mammoth job of making a road beside the long-ago surveyed line which marked our southern boundary and promised access to the Development road. This cut off as much as twenty miles on a trip to town as we had previously had to go via one or another of our neighbours' properties.

The "dozed line", as we called it, crossed a five-mile-wide band of melon-hole country which ran along the whole of our frontage to the road. The melon holes were depressions from four to ten feet deep and varying from ten feet across to big enough, in Joe's words, to float the Queen Mary. No one seemed to know how they had been formed and we reasoned that they were probably so named from the outback cattleman's habit of scattering pumpkin and melon seeds around during the dry season and reaping the harvest after the Wet. The water in the melon holes would ensure a good crop.

Because of his powers of observation I often called Joe "Old Eagle Eye" and he was true to form finding that surveyed line through the thick timber which had overgrown it. When his road was finally made roller-coaster up and down through the melon holes, it took twenty minutes to drive the five miles — on a good day, that is. After rain you didn't attempt it at all and in the windy months it was mandatory to carry a chainsaw to remove fallen trees from the track. But to us it was a great leap forward; it represented quicker access to the main road and it meant that it was now possible to fence the southern boundary.

The Federal Government announced a $20 million rural loans allocation just two weeks before the Queensland and Territory elections. It was seen in the bush as a cynical bribery attempt and it didn't work. About the same time we got a letter from the Development Bank, knocking back our application. A lot of people must have voted the same way we did. In Queensland, Labor lost 12 seats, leaving them with too few members to form a Shadow Cabinet, and in the Northern Territory they lost the lot so there was no opposition at all. Such results ought to have carried a message to the Federal Government but the Prime Minister, Gough Whitlam, continued to address public meetings of farmers, telling them that "they had never had it so good!" The mood of militancy in the bush grew apace, still largely ignored in the urban press.

On the other side of the gloomy coin, early storms gave promise of a good Wet, Bob's arduous studies were over when he sat for his last Senior exams and Kim went to Townsville for the wedding of her two university friends, Nell and Andy, who had both visited Cattle Camp a number of times and were already accepted as part of the family. (Being accepted as part of the family meant no frills, bringing your own swag and standing in line when the jobs were handed out. Nell and Andy are still doing just that!)

Thanks to the Utah Coal Company's policy of offering vacation employment to local upper high school and university students, Bob and Roger Chaffey from Seloh Nolem, as Grade 12 graduates, applied and were accepted for holiday jobs. The pay and conditions were extremely generous but the boys' first day was an eye-opener for them. Allocated to the same crew, they were standing waiting for instructions when a truck loaded with heavy bags pulled up beside them. The driver had obviously been driving all night and when he climbed wearily from the cabin and onto the tray and began to unload, the boys immediately jumped up to help him. It was second nature to them; if they hadn't automatically done that on any bush job, they'd have been bellowed at and told they weren't worth feeding.

A roar of voices stopped them in their tracks. "Demarcation!" "Twenty-four hour strike!" The foreman explained Union policy to the puzzled boys and after some persuasion the miners decided to overlook the incident just this once. Everyone sat and watched while the exhausted truckie unloaded six tonnes of bags all by himself.

"How the other half lives!" said Joe when Bob related the incident. We were so disgusted that we were never able to give much credence to the union myth of the poor, downtrodden workers — not on the Queensland coalfields anyway. They had over-award pay, bonuses, free work clothes, house rental at $5 a week and in the single men's quarters Bob paid only $10 a week for full board and lodging and the Company gave the married employees a large turkey for Christmas. (Cattle Camp scored two of those turkeys that year — passed on to us by miners who were going away for Christmas.)

Kim returned from Townsville with Tracey and Jimmy just before the bushfire which swept through most of our frontage country. When Joe noticed the smoke rolling in, he left the dozer in the area which he had just cleared of debris and ran two miles back to where he had parked the vehicle, just in time to drive it clear of the approaching flames. For once the Landcruiser started on cue and didn't have to be cranked. When I said I was relieved that it hadn't played up, Joe said not half as relieved as he was.

There hadn't yet been any run-off from the rain showers so the little dam near the bore was now boggy. Although the trough was a mere hundred yards away, one old cow persisted in trying to drink at the dam. It became a regular daily chore to drive down and pull Mad Cow, as we called her, out of the mud. It was a red-letter day, worthy of inclusion in the diary, when we finally saw Mad Cow drinking at the trough in mid-January.

"Well, naturally," said Joe. "She can't get to the dam, can she, because the creek's running a banker!"

A few days later Mad Cow swam the creek to stand in her usual drinking place, even though the water level was now up to her brisket.

That year looked like being one of the good Christmases. Bob passed his Senior subjects and Tracey won first prize in the Queensland School Students' Short Story competition for grades nine and ten, which she had also won the previous year in grade eight. On Christmas Eve we all went over to Seloh Nolem to celebrate with the Chaffeys and pick up Bob, who had ridden pillion on Roger's motor-bike across-country from the mine site.

It was midnight before we got home but Joe was up early to do the fence run and rescue Mad Cow before we all gathered for the present-opening ritual by the Christmas tree. This year the affluent wage-earners had bought presents and the results of the secretive work of the previous days by Kim, Tracey and Jim were revealed. Kim had made beautiful leatherwork handbags and belts, Tracey's shirts and blouses were a dressmaker's pride and Jim had painted each of us a small landscape in oils. The turkey and frills dinner was a far cry from last year's goanna repast. "No bushfires, no floods, no drama this year," said Joe contentedly as the girls began to clear the dinner table and I idly switched on the radio for some Christmas music.

Instead, we heard the urgent announcement of the devastation of Darwin by Cyclone Tracey. We were appalled. My sister Helen and her family lived only two streets from the sea at Nightcliff, which had born the full brunt of the blow. Nearly fifty deaths had already been reported; seven ships were aground in the harbour and six more were missing at sea.

I had received a letter from Helen only a few days earlier in which she had delightedly proclaimed the additions, repainting and new furnishings for the house, the culmination of years of determined saving. There was even a new bricked-in workshop for Ray under the high-set tropical home.

For days I monitored every news flash and periodically tuned in to the Alice Springs Flying Doctor Base where Dave, the operator, was checking the flood of refugees for me to no avail. It was New Year's Eve before we finally got a telegram from Perth. Helen's university student daughter, Dianne, had arrived in Darwin by plane only hours before the cyclone struck. She, with an injured leg, had been evacuated to Perth with the two younger children. Helen, Ray and Steve were camping in the wreckage of their home. They had sheltered in the workshop, with only its walls left standing. I howled in my relief and vowed to Joe that I'd never complain of my lot again.

Chapter 22

The Beef Depression Continues.

Bombandy had been pioneered by Alistair Brown's father almost a century before, but Alistair and his wife were childless and now Alistair was in his mid-seventies and the property was up for sale. I always thought of Alistair as the best example I knew of the outback term "a bit of a death adder" — to describe men so shy of strangers that they dodged human contact whereever possible and never left their home territory unless absolutely necessary. When we first came to Queensland Alistair had not been off the property for five years. His last trip to town had been to buy a short-wheel-based Landrover, when he had ended up buying four at once so he wouldn't have to repeat the trip too soon. Two were already on his scrapheap near our boundary; Alistair demanded the same performance from them as he expected from a horse, which meant that they were driven cross-country more often than on a track and consequently had a shortened life expectancy.

Now Bombandy was sold and Alistair and Mrs Brown had retired to the small country town of Clermont. The new owners were the Laidwigs with a daughter and four strapping sons, so Jim Somerville's services were redundant. Jim had worked on Bombandy for fifteen years and he asked Alistair, on leaving, if he could buy two favourite horses he had broken in some years before. His request was refused. I recalled that when Malley had left Mongrel Downs to take a manager's job, Joe had given him six of his favourites.

Jim was well known in the district but no one could afford a permanent employee so he took on various short-term contract jobs and was never at a loose end. He asked if he could leave his accumulated personal belongings at Cattle Camp so he could travel light and Joe was happy to agree because Jim's advice and detailed knowledge of the country had been such a great help to us when we were new chums.

We began the year with a spell of those breakdowns which so often seem to go in threes. First, the old stocktank sprang a leak, or rather a cascade, and it took a couple of days to patch the inside with cement and chicken wire. No sooner was that mended than the ancient pumpjack ground to a halt so there was no water to refill the tank. It required another two days and all Joe's ingenuity to get that going again. Then we had a week of switch mutiny. The switch on the transceiver played up and Joe had barely fixed that when the radio on-switch wouldn't work. He bypassed that switch with a couple of wires but he couldn't use that technique on the Toyota. It switched on okay but adamantly refused to switch off when the vehicle stopped. Joe had to leap out each time, throw up the bonnet and rip off the battery leads.

Instead of my usual sympathetic concern I derived a certain mean satisfaction from watching this procedure because I was crooked on Joe for some forgotten reason. For days we had only addressed each other in monosyllables.

"It's Mum's fault," said Tracey darkly. "She's a witch! I know, because I can do a bit of it myself!"

"You could be right!" agreed Joe.

At smoko time he came in with a broad grin and a bunch of wildflowers tied together with tie-wire and slid an arm round my waist. I burst out laughing. Next trip the switch behaved perfectly. But the vehicle suffered other breakdown problems they couldn't pin on me, so much so that my weekend transport from the Highway could no longer be guaranteed.

If the bank didn't consider us a viable prospect, the Teachers' Credit Union had no such reservations; they gave me an immediate loan to buy a two-wheel-drive Hilux utility and docked my fortnightly pay cheque to pay for it. Joe applauded my vehicle choice.

"Apart from getting to and from school," he said, "it's just the thing to pick up a load of station gear in town when and if we ever have enough money to buy a load of station gear in town."

"And," added Kim, "Mum's not stupid. It's not a four-wheel-drive, so Dad can't borrow it and knock it about on the run!"

When Bob's vacation employment came to an end in early March he had saved enough to add a third vehicle to the Cattle Camp stable, a small second-hand Cortina. But not for long. With Kim's agreement to stay on as Joe's offsider until the end of the year, he sent a telegram to Malley enquiring about a stockcamp job back in the Territory. Malley's affirmative reply was prompt.

"Good show," enthused Joe. "And when the lad gets a bit too big for his boots, Malley will pull him into gear quick-smart! Couldn't do better than work for a top man like Malley!"

"And," I added thoughtfully, "Malley won't sometimes treat him as if he's still just a kid."

Joe gave me a long look.

"Yeah, that too," he admitted with a little twitch of the lips. "Subject closed."

Bob packed swag, tuckerbox, saddle and bridle into his little car, hugged us goodbye, and Joe escorted him to the main road in case the car got bogged.

"Buck up, Hon," he said when he returned. "He's eighteen and feeling his oats. High time he had a spell of being footloose and fancy free."

"He mightn't come home again," I whimpered. "After all, you didn't and I didn't."

"Well," said Joe philosophically, "that's life. And don't forget, Hon, you've still got me." Then he added mournfully, "Thrown and tied and yarded for life!" I threw the dishcloth at him and for once got him before he ducked, so that cheered me up a bit.

Joe worked extra hard that year. The cattle were regularly mustered and branded with Kim's and frequently Jim Somerville's help. Jim brought his branding iron with him and took his pay in calves, branded and left with their mothers to be sold for him at some future date when prices revived. The current prices were at an all-time low. There were many reports of cattle sales where the graziers didn't make enough on their consignments to pay for the cost of transporting them to the saleyards, thus not only losing their cattle but going further into debt to do it.

In August the "Country Hour" on the radio reported the statistic that half of Australia's cattlemen were not viable. Despite Elders' hints about debt reduction, Joe categorically refused to risk selling any more stock. We were in a better situation than most because our herd had not yet bred up to the numbers the country could run, while on many longer-established places calves were being knocked on the head at birth to try to keep the numbers down to manageable proportions.

Despite his refusal to sell, Joe somehow managed to persuade Elders to extend the debt to cover a consignment of fencing materials and some new troughing. He plugged on, week after week, fencing, building another turkey-nest tank, setting up a new watering point, with Kim offsiding where practical or track-riding and doing the maintenance chores. She made pocket money with her carved leather handbags and moccasins, and painted whenever time allowed.

School holidays were synonymous with full-time mustering on our property as well as all the other properties with school-age children but, increasingly, many children were obliged to leave school and work with their parents during the day and continue their studies by correspondence at night.

As cash ran out, barter came in. Neighbours helped each other to muster and brand and a side of beef might be swapped in town for urgent vehicle repairs or a surreptitious drum of petrol or any of a variety of necessary items. It was often illegal but it was always a case of survival.

In 1975 politics was the major subject of conversation in the bush; never before had the usually phlegmatic and conservative country people been so politically aware or so fiercely angry. Where some countries subsidised their beef sales, we had a 1.6 cents a pound levy imposed on ours which the government proposed to raise even higher. When, in August, a Government Research Committee suggested that a Minister for Radio should be appointed, the implications must have frightened city as well as country people. A government which recognised Russia's claims to the Baltic States and fraternised with PLO terrorists was one thing, but government-controlled radio was definitely not on in a true democracy.

It all came to a head on November 11th when Sir John Kerr dissolved the Parliament and called a re-election. There was joy and celebration in the bush on a scale I'd only witnessed once before — the day the Second World War ended in Europe.

I could never understand the later tirades against Sir John Kerr who was, after all, a lifelong Labor supporter and a Labor appointee. Surely the subsequent landslide election results confirmed that most Australians approved his action.

Bob came home in early December with his girlfriend, Linda, in time to make a small contribution to the election win. Elections often coincide with our wet season, so it is common practice for station people to apply for postal votes. The only trouble is that, because of mail cancellations due to rain, and roads closed to motor vehicles, it isn't always possible to return the votes in time. This was the case in 1975 and Bob made a hazardous but successful motor-bike trip to deliver the Cattle Camp votes and those of some of our neighbours to a polling booth in the nearest town.

All things considered, 1975 was ending a lot better than it had begun. We'd had five inches of rain in October and follow-on rains just when they were needed and the cattle were fat and shiny. With the family members all home in exuberant high spirits and bursting with energy for the routine work, Joe was able to devote time to a project he'd had in mind for some weeks.

In November he had asked to buy some scrap metal from the dump at one of the mines and was quoted a very reasonable price. He took his oxy-equipment one day and cut what he wanted to make up a load for Brian Moohin to cart as back-loading on a routine trip. Brian delivered the load to our road frontage and wouldn't accept any payment for the job. The mine never did send an account either.

Joe drew meticulous plans and measured and cut steel sections from the scrap metal he had ferried in, load by load, from the road frontage.

"What are you making?" I queried, surveying the puzzling heaps of steel pieces laid out on the ground at the back of the shed.

"Wait and see!" said Joe. He didn't like his concentration interrupted for explanations while he was working and it wasn't until he loaded the oxy-gear and the assorted pieces onto the Toyota and drove towards the yards that I began to get a clue. It took a couple of weeks before he finished and was ready to demonstrate. He had made a cattle bail to handle grown cattle for branding, dehorning, spaying or castrating, a piece of equipment that cost over a thousand dollars to buy new. I like to think I made my small contribution. I had supplied the free school chalk, rescued from my pockets on washdays, with which he marked his measurements on the steel lengths.

Christmas week came and with it the flying ants. They descended in such clouds on Christmas Eve that we had to extinguish every light and sit in darkness while Joe made a bonfire well away from the shed to attract them away. (There is no chore more frustrating than trying to sweep up bucketfuls of shed termite wings inside a dwelling.) But these flights are welcome, not for their wings, but for what they signify. They are the prelude to the setting in of the real wet season, with its guarantee of widespread soaking rains.

Again we had two turkeys, this time provided by Joe after a quick sortie to the Dozed Line where the birds gathered in some numbers. Joe took the day off on Christmas Day, his first day's holiday for the year.

As Linda was overdue to return home to Alice Springs, Bob decided, two days after Christmas, to brave the threatening flood rains further north to take her to Townsville to catch the bus. The normal two-day return trip took him eight days in all. It would not have been possible at all if he had waited any longer.

On New Year's Day we invited the Chaffeys and Jim Somerville to come over for dinner. Jim arrived mid-morning. I had told the Chaffeys about the new and shorter Dozed Line road made since their last visit and when they hadn't turned up by 1 p.m. we thought they must have got lost or broken down, so Joe and Jim went looking for them. The Chaffeys arrived ten minutes after their would-be rescuers had set out; they had come via the old track through Bombandy. Kim jumped on the motor bike to chase after Joe and Jim but didn't manage to catch up to them until they had got right out to the main road. The upshot was that we didn't sit down to dinner until three o'clock.

The girls had pushed two tables together and covered them with a white damask cloth and Tracey had set out the good silver cutlery and unearthed a branched candlestick with three candles for a centrepiece. The festive board quite lifted the tone of the corrugated iron walls and the cement floor. The only trouble was that all this had been done at midday and three hours of excessive January heat in the tin shed had wrought an astonishing change on the candles. The one closest to the wall facing the afternoon sun had softened and drooped over till its tip was almost touching the tablecloth; the second was bent over in a wider arc but with gravity already exerting its downward pull and the third was still erect for half its length bending from the waist, so to speak, in the effort to catch up and droop like its fellows. Jimmy wondered aloud whether they'd stand up straight again if we turned the stand around to face the cooler wall. The food was still in the fridge so that was okay.

The Chaffeys left early enough to get home in daylight, vehicle willing, which was a wise precaution because their Landrover was over twenty years old and even sicker than our Toyota.

Chapter 23

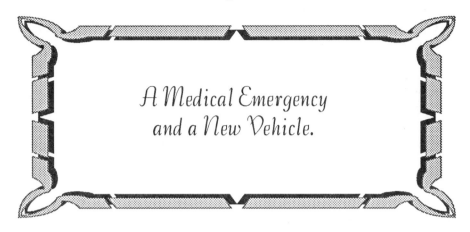

A Medical Emergency and a New Vehicle.

Cyclone season again. We were relatively blasé about cyclones because we were eighty miles inland and cyclone winds lose much of their velocity once they cross the coast. Our October hot-wind twisters were far more likely to unroof buildings and rip down trees and fences.

However, when Cyclone David crossed the coast in early January and headed directly our way, the reported wind speeds were still frightening. General Joe marshalled his forces.

"I'll check every nail and screw in the shed roof. Bob, you do the same with the little shed and then top up the water tanks so they're heavy enough to stay put. The rest of you, shift under cover every loose item the wind might lift!"

Clearing up a suburban backyard is one thing but a station homestead complex and scrap-heap, belonging to a family, none of whom rate highly for tidiness and some of whom pull assorted engines to bits and strew the ground with nuts and bolts and half-filled tins of dirty sump oil, is quite another.

By the time night fell, General Hawk-eye was finally satisfied and we were all utterly exhausted. We brought the dogs inside and waited apprehensively but by midnight all except Joe had fallen heavily asleep, so tired that we slept right through the roar of driving wind and belting rain as the cyclone passed over. In the morning Joe described the battering from one direction, the complete calm of the eye for twenty minutes and then the renewed assault from the opposite direction. The ground as far as we could see was covered with leafy branches and twigs torn from the trees and the rain gauge measured four inches.

Nobody took much notice of me when I asked plaintively who was going to shift everything outside again — there were engine parts and half-sheets of corrugated iron and cartons of junk and nearly-empty four-gallon drums. They all encroached on our living quarters. Joe and Kim happily cleared just enough space to set up their easels and paint for the next few rainy days while the others were detailed to check and mend fences.

I had the shed nearly to rights again by the time the school holidays ended. The rain periodically isolated the station but Bob managed a trip to town in late February and brought home the customary treat, a bag of cooked prawns, which Kim set out for lunch for Joe, Bob and herself. Joe had eaten only two when his hands and feet began to itch and burn. As the burning spread over his body he jumped under the shower and then staggered to his bed. Kim checked his temperature and pulse and called the Flying Doctor in Cairns immediately on the transceiver emergency channel, to report that both temperature and pulse were astronomically high. By this time Joe, though still conscious, was quite unable to move or speak. He said afterwards that he could clearly hear and understand all the exchanges between Kim and the doctor.

Quickly the doctor prescribed an injection of one of the numbered drugs in the Medical Chest. Bob raced to the airstrip to clear the cattle off in case a plane was needed for evacuation.

The doctor told Kim that if the pulse rate had not dropped near to normal in five minutes she must give an injection of adrenalin straight into the heart muscle and explained how to do it and stood by as she prepared the syringe.

"Now," said the doctor, "time's up. Take his pulse again."

The rate had dropped considerably but was still very high. Joe, conscious but paralysed, marvelled at Kim's calm efficiency.

"We'll wait another five minutes and check again," said the doctor.

The next reading was much lower and the danger past, although Joe was weak and shaky for some hours and his blood pressure did not return to normal for nearly a year. The doctor said that it was the worst case of sudden shellfish allergy he had ever encountered and it would have been fatal if we had not had the right Medical Chest drug on hand. Once again, we had good reason to say "Thank God for the Royal Flying Doctor Service!".

Kim left Cattle Camp in April to return to Alice Springs, her future now decided as an artist. Bob got a job at the Mines shortly afterwards and Fate decreed that Jimmy should take his turn as Joe's offsider.

Tracey wrote to tell us that some concerned boys in Jimmy's House had told her that he was being bullied unmercifully by three sadistic 16-year-olds who were repeating

Grade 10. Jimmy was 13, in the same class, and small for his age, but he had told no one, not even Tracey, what he had suffered nearly all term.

Joe was furious.

"He can come home for the rest of the year," he announced, "and I'll teach him how to handle himself with bullies!" Which he did. (By the end of that year I noticed that the lad had grown like a weed, toughened up considerably and was automatically referred to as Jim, instead of Jimmy.)

Although he worked hard for Joe during the week, he was mine for the weekend. I was disgusted with his low standard in French, so we started from the beginning and covered two years' work so he would be up to standard by the time he went back to school the next year to start Grade 10 again. He also repeated the Grade 9 Maths course.

With Tracey's holiday help, Joe and Jim mustered and drafted steers for sale.

"We'd better keep Elders happy," commented Joe.

Alan Bignell organised an on-the-property sale and we sold 140 steers for an average of $58 and Joe, Tracey and Jim helped to drove them to the Lotus Creek buyer's property.

That sale lowered our debt to Elders but didn't give us any cash in hand, so Alan helped us to prepare a loan application to the newly-formed State Rural Reconstruction Board. What a boost! It only took seven days for a low-interest loan of $25,000 to come through. Joe paid some outstanding debts, did some sums on the back of an envelope, and bought Jim a small Honda motor-bike, easier to ride round the fences than the unreliable bigger bike.

Our close friend, author Tom Ronan, died in Adelaide in August. I was home on school holidays when, late one night, we were awakened by the arrival of a vehicle. The visitors were Brian, Kevin and Kim Ronan, whom we had not seen since they were schoolboys holidaying on Mongrel Downs. They had taken their father's ashes to the family cemetery on Springvale Station in the Territory and had made the longish detour through Queensland on their way back to Adelaide. For the next four days every job that could be, was postponed as we reminisced and filled in the intervening years. Tom would have been proud that the art of vibrant conversation and yarn-telling, at which he was so proficient, was alive and well in his sons, though he would have been first to admit that Joe more than held his own.

Joe expressed himself almost as colourfully a couple of months later when I asked him what was wrong with the Toyota. I had found it abandoned half-way down the Dozed Line on my Friday return home, jacked-up with an axle and other assorted components lying on the ground beside it.

Joe had walked home the previous Tuesday and after a lengthy post-mortem the next day had verified his fears that the old Toyota was indubitably dead, had scrub-bashed its last anthill, had outwitted its last spewy bog. His only remaining mode of transport was horse or sick motor-bike, neither of which could carry a load of barbed wire or a roll of poly pipe.

The only other ---------!

"No!" I said firmly, now too spoilt to relish the thought of hitch-hiking again. "I'll phone Elders on Monday and get us a new one!"

"The loan money's all bespoke," said Joe, "and I doubt Elders will come to the party seeing I told them I won't sell any more cattle till the price comes good!"

"No harm in trying," I said.

When I phoned Alan Bignell and explained the situation his first question was "Will you sell some steers to cover it?"

"Yeah, sure," I said.

I explained that Joe had no way of getting to town to pick up the new vehicle so Alan arranged for its prompt but unexpected delivery a few days later. My surprised husband's delight knew no bounds when it arrived. If anyone deserved a sound vehicle which didn't have to be cranked for twenty minutes every morning before it would start, my Joe did. When I got home on Friday he was still in raptures.

I was glad I wasn't home during the next week when Alan called to discuss selling the cattle. Joe hit the roof.

"I'm not selling any cattle," he roared. "I thought I made it perfectly clear I wouldn't. I'll go without a vehicle rather than practically give 'em away! You can take it back! Marie's not the manager; she's got no say in when we sell. You oughta know that!"

Well, Alan didn't take it back, but it was years before Elders would trust me again.

Chapter 24

Holidays and House Plans.

Christmas 1976. Tracey had marked the end of her school career by once again winning the Short Story Competition, this time for Grade 12. Journalism at university? A cadetship on a newspaper? No way. Tracey planned to go back to the Territory to become a horse-breaker.

We had a transceiver schedule with Kim on Christmas Day. She was at Ross River, employed to take American tourists on early-morning horse rides, but with the rest of the day free to paint. Her paintings were selling well and she had saved nearly enough for a fare to England.

In the New Year the newly-formed Cattlemen's Union held a meeting at nearby Carfax Station, which Joe attended. There was no love lost between the producers and the meat-processors, who were holding down prices by strategically flooding the market at local sales with cattle from properties they had bought cheaply from families forced to sell up. It was decided to start a co-operative meatworks in Rockhampton. A small meatworks was available for purchase and those members who couldn't pay cash for the $100 shares could pay in kind, three bullocks equal to one share. Joe and I agreed that it was time the producers fought back.

Tracey and Jim made a good job of their holiday project — digging postholes for the extensions to the yards with crowbar and shovel. Jim had grown and filled out and was bent on proving that anything Bob could do he could do as well or better.

"Bob's a good man on a crowbar!" Joe commented slyly at the meal table. "He might give you a hand if he comes home next weekend." Bob now had a permanent job at the mines. Jim sweated all the harder. So did Tracey, already a budding feminist out to prove that she was the equal of any boy at any job.

Joe couldn't help them because he was nursing an injured shoulder and leg. While mustering a week before a big mickey bull had charged his horse. The horse reared in fright and toppled over, pinning Joe's leg to the ground. The bull couldn't stop in mid-charge and tripped over them both, stamping on Joe's shoulder as it came down. No bones were broken but he was considerably bruised.

We had decided that Jim should finish his schooling at Moranbah High School. I was impressed by the high teaching standards of my co-teachers and although he couldn't live with me in my supplied accommodation I was able to find weekly board with various friends for the next three years. Joe found his weekend assistance invaluable.

Kim came home for a month in April before leaving for London and Tracey took a jillaroo's job making her way back to the Territory. Cattle prices did not improve much and we sailed very close to the wind to maintain viability. That was the year we first employed Bill and Gloria McHugh on contract to help Joe muster. They had a great reputation for cleaning up properties where wild cattle had never been mustered before and were known as "the bull-tossers". Bill was our age and a dyed-in-the-wool cattleman with a fount of colourful bush yarns and Gloria, ten years younger, slim and wiry, was, in Joe's estimation, the best ringer with the smartest horse and cattle sense that he'd ever encountered. They were soon to become our firm friends for life.

With their help some of the big cleanskin mickey bulls that had previously dodged us in the brigalow got their comeuppance. With a stout tree lined up, Gloria lassoed them from her horse, then jumped off and dodged the bull round the tree until it wound itself up tight, while Bill stood by to fasten the rope. The last job each day was to go and pick up the tethered mickeys with a contraption Joe made that we called "the sled". The by-now chastened mickey was wrestled and tied down on the sled (and sometimes sat upon by the kids to make sure he didn't get loose) and towed back to the yards. Worse things happened to him in the yards next day, so that he was no more a virile young bull and might even grow into a saleable bullock. Whatever else, he wasn't going to steal the prerogative of our legitimate sires any longer and father nondescript calves.

The day when Gloria inadvertently caught Bill too because the roped bull swung wide he very naturally yelled, "Let go the rope! Let go!" as he was pinned against the tree. She didn't.

"If I had," explained Gloria later," the bull would have charged my horse. I'd tied it to a tree close by and it wouldn't have been able to dodge those horns.

Bill had a better chance, from what I could see."

It was a fair while before Bill could see any humour in the incident.

Once the full muster was over, Joe drafted out 167 young steers for sale to our buyer of the previous year at $50 a head and sent the culled cows to the Rockhampton sale, where they realised a much better price than we had expected.

A new pattern to our lives began to take shape. Bill and Gloria came regularly each year as contract musterers. For the next three years Jim worked with Joe each weekend but spent his Christmas holidays at the mines as part of the student employment scheme.

I was never really happy about Joe working alone for five days of most weeks. There were so many ways in which accidents could happen with horses, tractors or chain saws. As we neared home on Friday nights I always held my breath until I saw that the light was on in the shed and knew that all was well.

"Hon, you could break a leg in a buster from a horse on Monday, miles from home, and lie there all week for all we knew!"

"No, he couldn't, Mum," said Jim, "the wild pigs'd get him long before that!"

"Don't worry, Hon," Joe consoled me. "I'm always careful. I save all the risky jobs till Jim's with me on the weekends."

But I did worry; I couldn't help it.

During those three years both Joe and I managed a holiday. He was long overdue for a break, but couldn't be persuaded to take it and got snaky whenever I mentioned it, which only proved to me how much he needed it.

At the end of 1977 the mine job for the younger Grade 10 students only lasted until Christmas Eve, so Jim would have January home. I was flush with my holiday pay cheque so I cashed it and the first thing I did was to book and pay for a return air-fare to Sydney for Joe for two weeks in January.

Right up until the last minute he said he wouldn't go; if I was stupid enough to waste all that money that was my look-out! Of course, he capitulated and had a great time with his brothers and old friends, as I knew he would.

I met him at the airport on his return, looking fitter and more energetic than he had for ages. This was probably just as well because it rained heavily at home and Jim met us with the Toyota to escort us up the Dozed Line in case we got bogged. We did, and then Jim discovered he'd forgotten the tow-rope, so Joe had to put in some solid exercise on the long-handled shovel.

My holiday a year later was quite unplanned. I mentioned one weekend that some town friends with four children, who had been saving to take the family to Disneyland, had suddenly decided to borrow a large sum from a Credit Union and go right away before the children got too old to appreciate it.

"On a holiday," I said. "All that money just for a holiday. Why it's just as irresponsible as if I borrowed from the Credit Union to go to England!"

"Well?" said Joe.

"Well?" said Bob.

"Well, why don't you?" said Jim.

I looked at them in astonishment.

"You've paid off the ute," said Joe, "All your life you've wanted to go overseas and there's no way you'll ever get me outside Australia, so why not go while Kim's over there to show you around!"

"Oh, don't be silly, I couldn't!" I said.

"Of course you can, Hon!" said Joe. "If you don't book your ticket yourself, I'll do it for you!"

I argued that if I was going to borrow more money there were plenty of better things to spend it on.

"No, there aren't," reasoned Joe. "They wouldn't lend you enough for a tractor or a house and there's nothing else we need."

I went. I had Christmas with Kim in London and we went to Paris for New Year and got snowed in for a week in a cottage in Cornwall. It was like a dream; it was wonderful. I salved my conscience somewhat about spending all that money on myself when I took four radio scripts about outback Australia to the BBC and three were accepted at a good price.

When I got home the accumulated mail held another surprise. A newly established national farm newspaper to which I'd sent an article just on spec (I really was keen to earn that fare) had accepted it and invited me to write a regular column. Here I must explain that I was not to be an authoritative columnist on things agricultural. The situation in rural Australia was still so crook that the editor wanted some light relief among all the depressing news and my writing was so chauvinistically one-eyed, politically naive, and anecdotally exaggerated that it made readers laugh. The editor didn't say so in exactly those words, but that was what he meant. I liked to think that it was something like the light relief of the grave-diggers in Hamlet or the porter in Macbeth so I could lay claim to some literary merit.

That was when the house moved up the priorities list.

"Look," said Joe, "Any cleanskins we get are a bonus so they can go towards the building materials for a house. You keep a separate bankbook for what you earn writing, and that can buy some flash furniture and the trimmings."

I knew that he meant to set aside a proportion of future cattle sale proceeds towards the house but from then onwards we always referred to it as the *House of the Thousand Cleanskins.*

Chapter 25

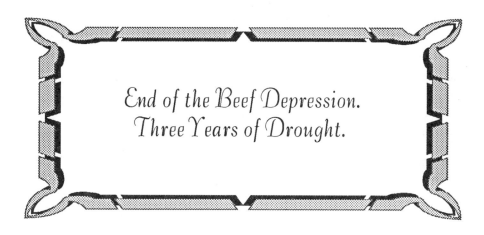

End of the Beef Depression.
Three Years of Drought.

We used to measure the agents' and tradespeople's faith in our eventual survival by whether or not we got a calendar from them at Christmas. We never did for all the years of the Beef Depression and had to depend on the newspaper calendar printed in the *North Queensland Register* every New Year.

When, towards the end of 1978, our Lotus Creek buyer paid $78 for the steers and then, in the following May, we averaged $215 a head for 54 cattle, Joe reckoned we ought to be good for a calendar from Elders that year. (It never rains but it pours; altogether we got seven calendars, two stock-agents' notebooks and three logo biro pens.)

Tracey returned from her year in the Territory, the horse fever almost worked out of her system. For some months she off-sided for Joe and we all missed her infectious high spirits and bubbly personality when she finally recognised her true metier and left home to begin her career in journalism. Meanwhile, Bob had made a friend at the mines who turned out to be the brother of Kim's best friend, Nell Harris. Paul came home with Bob nearly every weekend and soon became one of the family. His expertise as a fitter and turner was soon tested on our recalcitrant engines and sometimes uncooperative bulldozer.

That was a happy time with high-spirited young people home so often. With the seemingly endless energy of youth they tackled station jobs and sporting diversions with equal enthusiasm. I steeled myself to watch without protest when Bob and Paul took up sky-diving after Bob told me it was a substitute for rugby because I'd nagged him so often about playing such a dangerous game. I wasn't prepared to risk whatever might be considered a substitute for sky-diving.

I did, however, pull rank with Jim when he and Bob competed in both Alpha and Blair Athol rodeos one long weekend. Jim drew horses both times which reared so high he pulled them over backwards and was lucky to jump clear. I banned competitive rodeos until he finished Grade 12 a few months later, on the grounds that a rodeo injury could easily prevent him from sitting for his exams.

How quiet Cattle Camp was only a short year later. Bob and Paul had joined Kim, Nell and Andy on a bicycle tour of Greece and a sojourn on an Israeli kibbutz; Tracey was working on a South Australian newspaper and Jim was studying engineering at the University of Queensland in Brisbane.

I knew the wheel had turned full circle when I glanced at our meal table. When we were first married I used to clear off one end of the table to set our two places. Books, office work, maybe a spanner or a screwdriver, often a piece of sewing, covered the rest of the table. Now, after twenty-five years of being just a meal table, it was sneaking back to its old habits.

With fewer noisy human inhabitants around, the wild-life moved closer to the shed environs. Small grey wallabies and inquisitive emus grazed close by and Big Charlie, an old-man kangaroo taller than Joe, napped every afternoon under the bush-lemon tree a few yards from the back door. Mad Bat flew nightly inside the shed and, just outside, the little bettongs snarled and fought over their reclaimed territories. In the adjacent wattle trees the sugar gliders ran along the branches seeking gum bubbles for all the world like little grey squirrels minus the bushy tails.

Joe would never allow roo shooters on the place, saying that if there wasn't room for us and the roos too then it was time for us to leave. When I heard strange coughing grunts and snarls one afternoon I looked out from the shed to witness a three-way fight between Big Charlie and two other big roos. They charged and grunted and feinted blows at each other until Charlie finally sent them about their business. A few nights later I stayed up late reading, so I had to walk over to the lighting plant shed to put the engine off, a chore that Joe usually performed. On the way back I gave a couple of little coughs and the next minute heard a savage grunt in reply. I froze and swung the torch around. In its beam I saw Big Charlie balanced on his tail with his paws up in the classic boxing stance. "It's only me, Charlie," I yelled as I ran for the house gate.

I was very careful after that not to cough when I had occasion to put the engine off.

Life was good. The word "viable" no longer had the power to knot our stomachs and we felt confident that we could service our debt, given no untoward circumstances. The State Government's moratorium on the land payments during the Beef Depression had been a life-saver and a real morale-booster for the brigalow settlers. The re-writing of the payment contract at the end of the Depression with low interest rates over ten years was a further indication that the State Government of the day really did understand the rural economy.

We felt that we had weathered the storm. We had faced and survived most of the challenges that confront Australian graziers sooner or later and, in outback terminology, we had crossed our dry creeks safely and were battling on in good heart. But we had forgotten one thing, perhaps the most challenging of all — an extended drought.

The three drought years of 1980 to 1982 really set us back. They were the most depressing years of all and Joe's hair greyed suddenly with the worries of starving stock and the ever-present fear that the bore, our only water supply when the dams and waterholes dried up, might pump dry from the consistent 22 hours-a-day demand made upon it.

1979 had been a good year all round. Cattle prices had returned to normal and the season had smiled on us. The country we had cleared and seeded was a sea of waving buffel grass and Joe took pleasure in showing visiting friends around the property on his graded tracks. There were picnics on grassy banks by tree-shaded waterholes where we could swim or fish and, on the open, meadow-like tract near the big swamp, the brolgas and kangaroos mingled with the contentedly grazing cattle. The small tree orchids spilled in profusion from dead tree branches and everywhere there was the colour and cacophony of exuberant bird life.

"This is the icing on the damper, Hon," Joe would say. "The wild-life's happy, the cattle are happy and I'm happy too."

Sometimes he made sketches and colour-notes of trees and sunsets and sometimes he pored over pencilled plans and talked of sub-dividing paddocks and putting in laneways for easier mustering.

"But, first of all, we must have a more reliable water supply," he said. "Just as soon as we can afford it, we'll put a dam wall across Cattle Camp Gully. A water there would back up so far we'd never have to worry again. A pipeline up to the house too, so you can have all the lawns and gardens you want!"

How we dreamed on paper in the sharp winter evenings in the shed, crouching closer to the little kerosene heater, as Joe sketched and resketched the plan for The House of the Thousand Cleanskins. "An office for you, Hon, and a studio where I can paint, and four bedrooms because the kids will be home off and on, and visitors too. I'll build it in my spare time, shouldn't take more than a couple of years."

There was little spare time once the drought had set in, but we didn't envisage that it would last so long.

When the rains failed for the third year, to avoid further strain on the drought-stressed cattle Joe would not muster more than was absolutely necessary. It was enough to take the weaners off their mothers and to shift the cattle away from the waterholes and dams as they dried back and became boggy.

Joe worked under duress to fence off the waterholes which were now death-traps to weakened cattle and, at the same time, keep the water up at the only two remaining watering points, the troughs at the bore and the Turkey-Nest yards. Every engine on the place needed constant maintenance and Joe made sure that he always had a back-up YB engine ready for urgent replacement if the pumpjack engine should break down, which it did more than once. Even so, when the hot weather came and the cattle drank more, he had to give them a dry day periodically so the water could fill up again in the stock tank.

In 1982 I transferred to the new Dysart High School, half an hour's drive closer to home. I wasn't really sorry to leave Moranbah School. The Teachers' Union had suddenly become very militant and a group of its strongest supporters had joined the Moranbah staff and organised half a dozen strikes including one a week long, during 1981.

A few of us wouldn't join the strikers. The union then began a campaign against private schools with quite vitriolic propaganda, demanding the closure of these schools so all children had access to an equal education. Either they were ignorant of the fact that private schools were the only boarding schools, and that boarding schools were the only way, apart from limited-subject correspondence schooling, that bush children could get a secondary education; or else they knew but didn't care. Equal access only meant equal for children who lived in the cities and larger towns.

That was enough to prompt my resignation from the union. And didn't they get stroppy! The hard core at school swung into action. They read aloud the union directive to the whole staff. No one was to speak to me or collaborate in any way. Anyone who broke the coventry ban on me would merit similar treatment. I wasn't unduly perturbed; actually I felt rather embarrassed for them at first, acting like a bunch of sulky kids. But when the organiser told me they'd picket my vehicle at the school gates during the week-long strike, I was sufficiently provoked to reply,

"You do, you stinker, and you'll get a nudge with a bull-bar you won't forget in a hurry!" They didn't picket.

Joe offered to come to town and thump that same ringleader after he threatened to slash my tyres and abused me publicly in the town shopping centre, but I declined the offer. I wasn't really intimidated.

When nobody replied to my cheery "good morning" in my staff-room, and I had finally located my chair that they hid daily, I delighted in a monologue of provocative and often scurrilous comments about unionism to which they couldn't reply because of their ban. Apart from myself, they were all men in that staff-room and in the early days one older man had defied the ban and replied to my morning greeting but they gave him such a torrid time that the headmaster had organised his transfer to another town. None of the men fraternised with me but all the women in the other staff room, who were not militants or married to militants, had more guts and chatted with me as usual. It was a dreadful atmosphere throughout the whole school and, as I was the cause of it, it was time for me to go. I applied for a transfer.

I admit to being a bit apprehensive on my first day at Dysart. I needn't have been. There I met the friendliest co-teachers and one of the only two headmasters I have known whose integrity was unassailable. The friendship and cooperation I received at school during the week did much to help me face up to the worsening problems of the drought which confronted me at weekends.

I honestly don't know how Joe stuck it out with the continual breakdowns of vital equipment, the ever present stink of dead cattle and the wispy grass and dying trees in the bare paddocks. Eventually he opened up all the paddock gates to let the cattle forage at will, so the strongest would survive because they could walk further to graze and would not be constrained by fences. He bought what hay was available to handfeed the horses and a nucleus of the breeding herd and it was pitiful to see those pathetic cows gallop frantically towards the vehicle when I helped to throw off the hay bales at the weekend. We had to ration them, just enough to keep them alive, because the hay was so hard to get, with ten times more needed in the district than was available.

"Buck up, Hon," said Joe with a light touch on my shoulder, when he found me howling one day over the death of a cow, one I'd petted and fed by hand and looked for particularly each weekend because she'd been a hand-raised poddy calf a few years before. It seemed like the last straw to me.

I was instantly appalled at my weakness. It was I who should have been supporting Joe, not making things harder. He was the one who had to do the thankless and repetitive work, to make the hard decisions and face the dreadful task almost daily of shooting those poor beasts that were never going to struggle to their feet again. He faced the stark drought every day without respite, alone for most of the time, and I had five days of every week among green lawns and spinning sprinklers and friendly company and the distraction of a job that took my mind off the misery of the drought. I never howled again.

Of course, it wasn't all drought and despair. In those years we had many young cosmopolitan visitors as a result of Kim's and Bob's time on the kibbutz. They had exchanged addresses with other volunteers, some of whom they later visited in their European homes and some of whom included Australia in their later travels. If Bob or Kim weren't home when their friends arrived unexpectedly at Cattle Camp, Joe enjoyed entertaining them himself. In 1981 Bob bought and adapted a small secondhand furniture van for a round-Australia trip with some of their closer friends — Bea and Dagmar, two German girls, Maggi from Sweden and Roland from Switzerland. The day before departure Paul, who had been helping Joe for some weeks, decided to join the travellers. Kim went too. The end result was Paul and Bea's wedding in January 1982.

Chapter 26

The Future Looks Bright Again.

When over four inches of rain fell on New Year's Day 1983, the worst was over. We were offered a good price for the station, walk-in — walk-out — with 1200 head of cattle, by an approach through Elders. This wasn't the first time we had been approached to sell over the years and the offer might have been tempting if it had come when our spirits were lowest during the drought. But now the icing was back on the damper and the world was turning green again. We rejected the offer out of hand and bought a new lighting plant to celebrate our renewed enthusiasm and guarantee consistent lighting at night time while we re-assessed our future plans.

The May flood rains finally broke the Australia-wide drought and consolidated our return to good seasons. Once again, the Dozed Line was impassable, under water for much of the distance. On Saturday mornings I drove the long way home on the bitumen to the mailbox, where I left the vehicle until my Sunday afternoon return. Joe met me there on the bulldozer towing a home-made carriage which transported the weekly supplies, my dog Tiger and me the rest of the way home. The two-wheeled carriage was really the back end of one of Alistair Browns's jettisoned short-wheel-based Landrovers and sometimes it almost floated in the bow wave from the dozer. It was a slow, two-hour journey each way and the mosquitoes and sandflies were savage and could only be foiled by waving a leafy switch backwards and forwards non-stop.

It was at that time that our whole district was shocked by a mindless and brutal murder. Only two weeks after the deluge, young Dave Parkinson, our next-door neighbour, was reported missing. His homestead was quite close to the bitumen road, and tracks showed that a car had turned in, got bogged near the house, been dug out and then returned to the road. Dave had been alone on the property as usual. His own vehicle with his dog still on the back was at the bog site as if he had gone to help rescue the trapped car.

For two weeks police and neighbours searched the property and the sides of nearby roads on horseback where it was still too wet for a vehicle to move. They were looking for signs of the disposal of a body because a blood-soaked patch of ground was obvious near Dave's vehicle. He was eventually found buried in a shallow grave in creek sand twenty miles north of his property, shot five times in the back and behind the ear. The police said it had the hallmarks of a contract killing.

This was not a very pleasant time for his neighbours, who were all interviewed at some length about occasional past disagreements, especially the neighbour who owned a car with the same tyre treads as the mystery tracks. Arrests were finally made when it was revealed that Dave had recently become engaged to the ex-mistress of an older city man who had withdrawn an unexplained $10,000 from his bank account a few days before the murder. Speculation was rife for months but the eventual trial was aborted because of conflicting evidence and the murderers have not yet been brought to justice.

The dozer and carriage ensemble was still our transport method when Tracey came home for a holiday for the first time in two years, which coincided with Bob's and Kim's return and Jim's mid-year University break. With the whole family, luggage and stores in the carriage, it was a merry if somewhat crowded trip from the mailbox to the shed. "A likely lot of builders' labourers!" said Joe as he surveyed them and announced the arrival of the building materials for the house as soon as the road was negotiable. He was right; all except Tracey, back at her interstate job, did help to build the *House of the Thousand Cleanskins* at some period of its construction.

The bettongs did not cede their territory willingly. When Joe put in the first two pegs and stretched a cord between them it was almost sunset. When he returned next morning, a bettong track led up to the middle of the cord, which had been chewed right through, ends limp on the ground. The little marsupial could easily have squeezed under or jumped over the cord. "I think he's making a statement!" said Joe. (They never did give up their land rights. Today, ten years later, we continue to co-exist. The yard is mine during daylight hours and theirs after dark. It is drought-time again and their natural food is scarce, so I feed chunks of bread to seven of them at the back step every evening. If I am late one of their number will hop up the two steps onto the verandah and peer enquiringly through the kitchen door.)

Joe's estimate of a couple of years to build the house proved correct. Kim and Bob helped to put the floor on the stumps, erect the walls and roof framework and put the roof on. Jim and I screwed down the verandah floorboards during our holidays.

When all work was periodically suspended for three or four weeks at a time while Joe, Bill and Gloria did the cattle-work, I quelled my impatience by day-dreaming about a future which included an indoor toilet, hot water on tap and no snakes inside the house. And don't forget the wild pigs. Like the one that charged in the open end of the shed and put me up on the table and the one that charged Joe at the back door, which he just got slammed behind him as the boar thudded into it. And the horses, come to that. They got into the shed more than once, trying to sneak the oats Joe kept there.

There were hold-ups, of course. A lot of the material had been short-supplied, which Joe didn't discover until he came to use it. With Brisbane so far away, it was usually at least a fortnight after our complaints that the shortfall arrived in Mackay and had to be collected there.

Another great leap forward hovered tantalisingly on the horizon. With the installation of an eight-channel repeater station to our north the connection of a radio-telephone was now possible. The only sticking point was the installation fee of $12,000.

"That's a bit rough!" I complained on a town telephone to the Telecom official concerned with the proposed scheme.

"You rich graziers can afford it!" he replied sneeringly and proceeded to rip into me about exploiting the rural workers while I lived in effete luxury on the sweat of their labour.

"Another union rep!" I mourned as I hung up the receiver on his tirade.

There was no way any of the settlers could raise that sort of money, so Telecom offered us an alternative. We could pay the installation fee at $500 annually, an amount they promised not to raise for three years. The catch was that the fee didn't cut out when the $12,000 was paid off; it would be a permanent extra charge on the phone and no saying to what heights the annual charge might rise.

"No thanks!" Joe said. The two dollars a year licence fee we paid for the transceiver was, on the face of it, rather better value for our contact, though limited, with the outside world.

When the settlers heard of a proposed visit of the relevant Federal Minister to Longreach we all contributed $20 to a protest group and two local wives drove to Longreach with our blessing, determined to see the Minister and put our case for fairer treatment. They were successful. The Minister heard them out, said he was quite unaware of the proposed charges and agreed that they were, indeed, a bit rough. As a result, we were offered and accepted a reduced installation fee of $1350.

Our radio-telephone tower was erected at the house site with a temporary line laid on the ground over to the shed where we still lived. When I got home on the weekend, Joe told me he was beginning to understand why the job was allegedly so costly. It took two vehicles and six men to do the job, most of which he could have done single-handed in a couple of hours. The truck with the steel pole and three stays had come all the way from Melbourne, costs all up about $6000, said the man in charge, because one Melbourne firm had the sole contract to supply the poles.

"Did they send the wrong ones then?" remarked Joe. "These aren't gold-plated. I could get the same thing for a couple of hundred dollars in Mackay!"

Two men took turns to dig the short trench for the cable as their only contribution; another two erected the pole and the battery cabinet and the last two tested the electronic components. (I have to be fair and say that nowadays Telecom does it more efficiently when a breakdown occurs; a single employee arrives promptly to replace any necessary parts.)

The phone was connected on the 28th of November, Bob's birthday, and the day before Tracey's. Apart from the birthday calls, there was one big advantage for me. If there were storms about I could now phone Joe on Friday before I left Dysart to find out whether the road home was okay, whether it was a bit sticky and he'd meet me at the mailbox, or whether it was raining hard and I'd better stay put and phone again in the morning. Previously, the time-lapse involved between Joe sending a telegram via Cairns on the transceiver and my receiving it meant that the road condition could have changed dramatically in the meantime and he had no way of up-dating the information before I left, or even of being sure that I had left, for a late storm at the Dysart end of the sticky black-soil road could easily prevent me doing so.

Telephone access also made it a great deal easier to organise the arrivals of family and friends at Christmas time. And what a wonderful Christmas we had, that last Christmas of all in the old shed. Everyone home: Kim from her art studio in Sydney; Tracey and her boyfriend driving all the way from Adelaide; Bob and his girlfriend from Moranbah and Jim from Brisbane; Paul, Bea and baby Jodan from Moranbah too. Nell and Andy came from Charters Towers and Lorna, Nell's and Paul's mother, from her Gold Coast home.

The wet season stores were safely bought and the old kero fridge laboured gamely to hold all the perishables. The visitors slept wherever they could find a bed or room to throw out a swag. There was the usual Christmas Eve ritual of the boys getting the Christmas tree and finding it too big to fit inside. "Get one the right size this year," I called after them as I called every year.

Then the try-out, too tall again, and the lopping to fit, and the girls decorating the tree, and the cooking of turkeys and the wondering whether the threatening thunderstorms would hold off until the cooking was done before the rain put the fire out in the outdoor stove. Periodically, Joe would quietly sneak off by himself for an hour or two on some vague job. After the year of mostly solitary days he had to come by degrees to cope with all the non-stop busy activity around him.

Christmas is always magic when the whole family and their close friends are together and there is an added quality to the magic if the first storms of the Wet have filled the waterholes and the spidery white flowers of the bush lilies are scattered in the green fringes of the swampy ground. These are not the Christmas lilies of the South, these smaller wild relations. They are common to all central and northern outback Australia, the desert as well as the savanna forests, and they come to short-lived life at any time of the year after the right sort of rain — enough rain to push away any fears of imminent drought. And that, to the outback family, is perhaps the best Christmas present of all.

That year the white lilies pushed up here and there along the Dozed Line two weeks before Christmas and it was a sure bet that soon their larger and grander blue lotus cousins would bedeck the waterholes in Scrubby Creek and bloom there resolutely until once more the waterholes dried back to inhospitable mud.

Chapter 27

*A Wedding and
a Really-Truly House to Live in.*

Although Cattle Camp is in a much higher rainfall belt than Mongrel Downs, when we lived in the desert we never had anything like the obsession about rainfall that we have in this comparatively lush country. There was never the worry about drinking water for the stock because that was supplied by ten bores which didn't depend on the rainfall. As for feed, the property was so large that, if there were storms about, you could be sure that at least some part was getting good grass-rain and, because there weren't any fences to stop them, the cattle could chase the storms to get a green pick.

At Cattle Camp, no bigger than a good-sized Territory horse-paddock, we often miss out altogether from storms which can drop two or three inches on a neighbouring property — and vice-versa, of course. Creeks and dams need good falls regularly to keep up the supply of drinking water for the stock, so we watch the cloud build-up anxiously every year as we never had to on the desert station.

In April 1985, the bore, our one stand-by, suddenly dropped its output to only 200 gallons an hour. This was worrying as the dread possibility of pumping it dry had to be considered. Luckily, the waterholes were brimming, so Joe laid a pipeline from the big Devlin waterhole to join into the Turkey-nest pipeline and pumped from the creek to give the bore a chance to make up.

He often had weekend help from the boys. Bob was working at the new Riverside Mine and Jim, first year out of university, was a cadet engineer at the Harrow Creek underground mine. Both mines were close to Moranbah and handy to home.

Meanwhile, closer to my own selfish consideration, was the water supply for the humans at the *House of the Thousand Cleanskins*. Joe cleaned out the small dam (on the site of the proposed big one), laid a pipeline to the house and organised the erection of an overhead tank. His next job was the house plumbing.

Kim came home for a month's holiday in September and teamed up with Joe, Bill and Gloria for a round of cattle work. She was home when Bob, on top of the world, introduced us to Liz Nott and announced their engagement. Liz was an art teacher at Moranbah High School. Her mother owned a property in the Banana district. (This location nearly got Bob into a fight a couple of years later, when he and Liz attended the crowded opening of The Stockman's Hall of Fame at Longreach and I met them there. In the evening the three of us found ourselves sharing a table at the pub with a trio of exuberant young fellows.

"Where are you boys from?" I enquired.

"Orange!" shouted one in the general uproar," all the way from Orange! Drove all day and all night!" and, turning to Bob,

"Where are you from, mate?"

"Banana," replied Bob, without thinking.

Three faces darkened, beer glasses mid-way to their lips.

"Smart-arse!" growled one, half rising to his feet.

"He really does come from Banana!" I interposed quickly. They subsided, but I'm not sure they believed me.)

The wedding was set for mid-December and Joe and I took a weekend off in November to drive down to meet Liz's mother at Darling Plain station, leaving Bob and Liz to hold the fort at Cattle Camp. Three inches of rain fell at home on Saturday night and Bob and Liz had the job of rescuing the pump and engine from the steep-banked Devlin waterhole with only their brand-new little Niva station wagon to do the job. (Joe had taken the Cattle Camp workhorse vehicle to meet me at Dysart, from where we had continued the journey in my more respectable Hilux.)

It wouldn't have been politic to remind the Niva's owners that they'd admonished me to wipe my feet carefully before I sat admiringly inside their spotless new vehicle early on Saturday morning. The mud inside and out on Sunday evening wasn't what attracted my attention so much as the water-level sloshing half-way up inside the headlight panels. "Good little bus!" said Joe, seemingly oblivious of its state, for it had indeed towed his engine up and out of the flood. I was glad Liz had a bush background and knew that incidents comparable to this were not uncommon in our way of life.

As the weeks passed quickly, Joe raced to get the house to the habitable stage. Last day of school. While Joe connected the water supply and the hot-water system I shot into town and spent my holiday pay on a fridge and freezer pair and a washing machine. We left early the next morning, Sunday the 15th, for the afternoon wedding in a lovely garden setting at Biloela, where the Settlers' Inn had been reserved almost exclusively for the wedding guests. Kim, Tracey, Nell, Andy and

Paul, were already there when we arrived. Helen and her daughter, Vanda, were soon to come.

The bride was breathtakingly beautiful, the threatening storm held off until the register was being signed and the reception was a true country celebration. Early next morning Bob and Liz left in a hire care for Rockhampton to catch a plane for their overseas honeymoon. They were going to spend Christmas in Switzerland with Bob's kibbutz friend, Roland, and his new bride.

With all the storms about we half-expected the Mackenzie River to be over the bridge on our way home next day. How we cheered when it was only lapping the planks. Kim, Tracey, Helen and Vanda were travelling with Jim in his car, a much better alternative than the back of our Hilux, at least until Jim realised he'd forgotten to fill up his petrol tank and would have to make a detour to Middlemount for fuel and lucky if he got that far. We waved unsympathetically to Jim. Helen squeezed in the front with us and the girls climbed on the back and so home to Cattle Camp and the *House of the Thousand Cleanskins.*

Proudly Joe lit the fire in the combustion stove and boiled the kettle for our first cup of tea in our new home. There was still a bit to do — verandah steps, internal doors and the wiring for the lights and power (not to mention the painting and floor coverings, but that could come later). Willing helpers spent the rest of the afternoon carrying things over from the shed — table, chairs, beds and bedding, crockery and selected items from the food cupboard and from the old kero fridge, while Joe put the finishing touches to the sewerage system.

We dined by lamplight that night, a merry meal, and the next day the electrician arrived to do the wiring and Joe dug the trench from the shed to extend the cable for the power. By the end of day two we had electric lights, verandah steps and a door to the bathroom. The first green frog had discovered the toilet bowl because there wasn't yet a door to the laundry and toilet which opened onto the back verandah (not that doors ever stop green frogs, as everybody in the bush knows full well).

Helen helped me plan a garden and envied my walk-in pantry and built-in linen press of marvellous dimensions. I was sad when she and Vanda left for Darwin a few days before Christmas. Jim came home the next day, bubbling with plans for a Tasmanian holiday after Christmas, and then we had an unexpected visit from Jim Somerville, who had left the district about three years earlier.

The Christmas tree had to be lopped even more than usual to fit inside the lounge and our pleasant family Christmas was over all too soon when Jim, Kim and Tracey left to drive south on Boxing Day. Good timing, though. That night, storm rain closed our road and every night for a week another storm followed. The Wet had set in.

The frogs chorused as loud as ever, but they were all outside. Only a few insects beat the barricades of flywired doors and windows. We didn't have to put on raincoats to go to the toilet.

So why did I feel a slight sneaky sense of disloyalty when I gazed across at the poor old deserted shed. I said as much to Joe. "Easy to understand," he said, sliding an arm round my shoulder. "We've lived in the shed almost twice as long as any other home we've had. Lot of memories there. I feel the same myself."

Then he reminded me of a Territory bachelor friend, who had a flash homestead built after many years of living hard on his property. He was very proud of it, but somehow he never did move in. When Joe called there six months after its completion, he was still living contentedly in his precise, neat two-room tin shed with its bough-shed verandah, where his three dogs flapped their tails as they lay, whenever they heard the sound of his voice.

"Been camped here a bit too long, I reckon, Joe," he explained. "Got everything just the way I like it. Besides, the dogs don't fancy the new place. Told me so. Got a flash homestead now like everyone else, but there's nothin' to say I gotta live in it!"

"And therein," I remarked, "lies a basic difference between male and female. There's not a woman alive, I'm certain, who wouldn't have been in that new house like a shot."

Then a new thought struck me.

"Would you have been happy to stay in the shed if I hadn't been here?" I asked curiously.

I knew, even as I asked, that I wasn't going to get a straight answer.

"A man works outside all day at hard, dirty jobs, he deserves a few amenities and comforts when he gets home," he replied.

"Oh, and here was me thinking you did all that extra work building a house just to keep me happy!" I said. "And wondering how I could show my appreciation."

He gave me that special, slowly broadening grin I loved so much.

"Well, I know how you could start!" he said. So I did.

Chapter 28

A Second Wedding.
Cyclone Charlie, the Dam-Buster.

There was still a great deal to be done inside the house before the painting could be attempted — bathroom and laundry tiling, built-in wardrobes, and the arduous task of plastering over all the nailheads and joins between the gyprock sheets of walls and ceilings. Hardest of all was finding the time to do it.

A much more urgent building program claimed Joe's attention and he switched from house-builder to yard-builder. A good set of solid timber cattle yards could take a team months to build and, if done by a contractor, might cost as much or more than the average house. We urgently needed a yard and loading ramp at the Turkey-nest but the $100,000 loan we had taken out with a mortgage to the Agricultural Bank of Queensland, large though it was, did not run to paying yard-builders to do the job. That was ear-marked for scrub-pulling the brigalow and melon-hole country and aerial-seeding it to improved pastures. This was one of the conditions, along with fencing and watering points, that we had to fulfil before we could get the title to the block when it was eventually paid off.

Joe planned the basic yards he needed, a combination of steel and local timber because he had to work with whatever materials he had on hand. Bob had earlier acquired a semi-load of second-hand eight-inch-diameter pipes from the mines and these Joe cut into lengths for posts for the main yard around the trough. Filled with cement they would last for ever. He used some piping for rails with temporary brigalow rails between them, and the race and loading ramp were also made of piping. He cut out the gates and slide gates from the scrap metal from his foray at the mines a few years before.

For some weeks Joe swapped hammer, drill and screwdriver for chainsaw, bulldozer and oxy-torch. The cattle, coming in to drink daily at the Turkey-nest trough, soon grew used to the human activity where once there had only been the sounds of nature.

In May, Tracey, working for an Adelaide paper, won the Australia-wide award for the best news-story of the year for weekly and provincial newspapers and gained her B-grade journalist's accreditation. As the runner-up story was also hers, we were very proud of her.

Towards the end of the year Jim brought his new girlfriend, Carolyn, home for the weekend and she charmed Joe immediately. We sat over a cup of tea after they had departed on Sunday afternoon.

"I don't know what's up with that lad!" Joe said. "Why hasn't he snapped her up? I wouldn't have been so slow off the mark!"

"They've only known each other a few weeks, Hon," I commented.

"No matter!" said Joe. "Lovely girl like that, sure to have all the fellows after her. Reckon we ought to kidnap her and keep her here out of circulation till he wakes up to himself?"

I said I thought we could safely leave it to Jim.

The main cattle muster was very late that year. Each time Joe set a date for Bill and Gloria to come, it either rained heavily a day or two before they were due or even the day after they arrived. Once it rained during the hour and a half it took them to drive between their Lotus Creek home and Cattle Camp. It was always enough rain to postpone the muster for sale cattle because the trucks would not be able to get to the yards to load them. Then, because Bill and Gloria were tied up with other contract work, they weren't available to come to us when the road had dried out enough for the trucks.

Contrary to the normal storm pattern, November was dry. Kim came home early and Jim took a week's holiday to help Joe trap a mob in the Homestead yards and draft 109 fats for sale. The work was in full swing when school broke up in December. It was hot and humid and tempers were a mite touchy because I don't believe any of us expected that the storms would hold off long enough to get the job done, calves branded and sale cattle away.

Bob and Liz called in for a couple of days before Christmas but only Kim and Jim were home on Christmas Day, our quietest Christmas yet. There were 800 head of cattle in the Homestead yards complex, some yards full of cows with bawling little calves. They all had to be fed hay and rotated through to the water yard daily to get a drink. Because it was so hot and dry the other watering points had to be checked regularly and Joe had to pump daily to the Turkey-Nest. Nobody was the least surprised when the pipeline sprang a leak, not once but twice, and had to be repaired. The lighting plant, sulky for weeks, began to play up in earnest.

Kim had to return to her Charters Towers studio on the last day of December and Jim left for the Sunshine Coast to meet Carolyn's family before his holiday was over. The engagement was official three weeks later. Joe, Bill, Gloria and I watched the clouds build up with our fingers crossed. On the 4th of January we got the sale mob away and, on the next couple of days, branded 312 calves, mothered them up, and then opened all the gates wide and let them go. Now it could rain all it wanted and the more the better; we never wanted to cut things so fine again.

But it didn't rain. A storm on the 24th saved us beginning water-rationing but then it continued dry right through February and we didn't get good rains until March. Joe took advantage of the dry weather and trapped another 60 head of fats for sale.

On the first weekend in March we had what was rather like a belated Christmas with all the family home. Tracey had driven her elderly little car all the way from Adelaide, amazingly without one hiccup, and, after learning night-school Spanish, was determined to set off on a freelance career in Central or South America. She had arranged to be a stringer for *The Australian* and *The Canberra Times*. We knew we hadn't a hope of dissuading her; she had already bought her plane ticket. Joe, suspecting that she probably didn't have a very healthy bank account, bought the little car from her for $1500. It would do very well, he said, to leave out near the mail box and the bitumen road during the wet season. If we covered the intervening distance from homestead to mailbox by tractor, horseback or on foot if necessary, it would then serve to take us to town or wherever we needed to go. As indeed it did, on more than one occasion. We left it hidden from the road by the trees and, as Tracey said, it was such an old model that nobody would be likely to consider it worth stealing.

In mid-year I took six weeks off from school without pay to accompany an American photographer, Melinda Berge, on a research trip for a book about the modern Australian stockman. It was Melinda's idea for an update for the coming Bicentenary, which she had submitted to The Stockmen's Hall of Fame and I had been asked to write the text. (I had collaborated in 1980 on *The Stockman* about the early days of the cattle industry in Australia.)

The publisher's contract included a deadline to submit the manuscript and, back at school teaching, I knew I didn't have enough spare time to do the work and meet the deadline. Before I left home on the Sunday afternoon with only a fortnight to go, I gave Joe a rough outline of what was needed for one of the chapters still to be written. He had read and edited what I had done so far and was quite familiar with my style of writing. (In fact, I'd often cribbed his observations and his turn of speech for my newspaper column, so it was really his style as much as mine.)

On the following Friday night I read what he had written for me and added only one comma; all the typing was finished by the next Tuesday and the completed manuscript posted on Wednesday. I'll bet nobody can pick out the chapter I didn't write.

"What's a mate for, Hon," said Joe when I thanked him.

The cattle work was done on time that year and the biggest percentage of the sale cattle trucked in September. Bob and Jim came on weekends to help brand the weaners.

We were able to pay off enough of our debt to Elders to keep them onside, with enough left over to build the big dam Joe had wanted so desperately for years. Too often we had lived on the edge of disaster; this dam should mean that we would never again be faced with the prospect of dire water shortage.

Joe sent our little bulldozer, workhorse of many years, for a complete overhaul and watched the contractor's huge earth-moving machines move in. After good rains a 400-yard wall across Cattle Camp Gully behind the dam hole would back up the water through the horse paddock for almost a mile. Joe estimated that it would probably take about four years' normal rains to fill the basin initially.

As happens more often than not, there were numerous hold-ups in the work because of machine breakdowns. In mid-November there had been widespread light storms and the country was the greenest it had been for years. By early December Joe began to feel that time was running out for the dam completion before the heavy rains came. The first wild storm silenced the lighting plant's complaints for ever when lightning blew a nearby bloodwood tree to pieces and leapt across to the plant.

Jim and Carolyn's wedding was planned for the first Sunday after school break-up, on the Sunshine Coast, so home problems were put on hold for a week. On the prior Thursday the northern contingent of the guests, Kim, Nell and Andy, Paul and Jodan, arrived at Cattle Camp to break their journey south. (Bea and the new baby were flying down.) The visitors helped Joe install a new lighting plant and left early on Friday. Joe handed over to Roger Chaffey to caretake for us and drove to Dysart to pick me up in Tracey's little car as soon as school let out, arranging to meet Kim at Middlemount so we could travel together.

We stayed at a Rockhampton motel that night and Kim got away half an hour earlier next morning while Joe and I stayed to do some shopping. It was about 11 o'clock and 20 miles south of Rockhampton when the little car suddenly changed engine note and stopped abruptly. Nothing, but nothing, could get the faintest squeak out of it, no matter what Joe tried. Long experience bade me keep my mouth shut as he probed the engine but I was certainly feeling desperate. I could hardly believe my eyes when Kim's little yellow truck pulled in behind us; it should have been well ahead. Luckily for us, she had taken a wrong turning out of town and it was some time before she'd discovered her mistake and retraced her way.

She drove back to Rockhampton for the RACQ breakdown mechanic but he couldn't isolate the trouble so towed the car back to his garage. How to continue our journey? There was only one spare bucket seat in Kim's truck. We knew the bridegroom was on the road somewhere but was he ahead or behind? It simply did not occur to us because we'd never moved in such elevated circles where such things occurred, but the RACQ man's suggestion was the obvious answer.

"Why don't you hire a car?" he said.

We hired a brand-new Ford Laser. What comfort!

"Hey!" I said. "Look, Hon, air-conditioning!"

Joe was a bit suspicious.

"I prefer fresh air myself," he commented. Well, I knew that because, on the rare occasions we stayed in motels, Joe always insisted on leaving the door and windows open all night. We compromised. He drove with his window open and I closed mine and put the air-conditioning on.

We caught up with Kim near Gladstone, but only reached Maryborough that night. It was mid-morning on Sunday when we arrived at the motel where rooms were reserved for the wedding guests. Once again Helen was right behind us after a circuitous trip from Darwin.

The bride was ethereally beautiful. Liz took the wedding photos, as Tracey had done for her and Bob's wedding, and at the magnificent reception Tracey's telegram from El Salvador was written in Spanish. Later, she told us that the postal officer had refused point-blank to accept it in English.

We all enjoyed the next day by the pool at Carolyn's parents' home and then Joe and I spent the next two days visiting old Territory friends who had retired in the area. Then it was back to Rockhampton, change back to our now-repaired little car, overnight at Liz's mother's property, then to Dysart for my Hilux and so home. Kim was already there, assuring our anxious pets that we really were coming back. She left for Charters Towers next day.

Bob and Liz joined us for Christmas, in time for Bob to help Joe complete the ringlock fence around the expansive house yard, which I fondly but wrongly believed might keep the bettongs out.

The dam wall was finished on the 14th of January, two days before Jim and Carolyn returned from their Norfolk Island honeymoon and Bob and Liz came from Moranbah. Joe proudly showed off the new dam and we all helped to broadcast couch grass seed in the red topsoil spread on the broad top of the dam wall.

It stayed dry until the last week of February and Joe had to pump constantly from the bore. It seemed as if the climate truly was changing; we hadn't had a real down-to-earth on-time wet season for years. About the only chance now of avoiding another drought would be a cyclone crossing the coast to the north and forming a rain depression.

We got one. Cyclone Charlie hovered off the coast for a week and the attendant light but steady rain filled the dam hole and began to back up into the horse paddock. On the 1st of March Charlie crossed the coast near Bowen and it really began to rain. Over sixteen inches in four days. The water rose within two feet of the top of the dam wall and poured in deluges round the bywashes at each end. It did not seem possible that the wall could hold; all over the country reports of ravished dams were flooding in. Joe checked anxiously every few hours.

Two days after the cessation of the downpour he saw the first trickle of water bubbling out at the back of the wall. A pipe had formed. When I phoned from school next morning he told me the news I'd been dreading.

"It's midway along the wall. The top of the wall hasn't collapsed in, but the water's pouring through a hole big enough to drive a Toyota through!"

Such is life on the land.

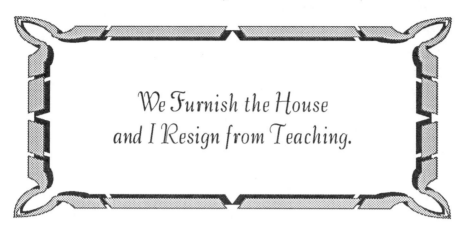

Chapter 29

We Furnish the House and I Resign from Teaching.

The bottom of the pipe was about three feet above the base of the dam wall so, when the water level stabilised, we still had a fair supply backed up. Joe hastily worked with the little dozer pushing dirt into the front of the hole. After a couple of days he had raised the dirt level about two feet higher and thick enough to cope with all but flood rain. The pipe had spread into a big cave in the back of the wall; this would be a major repair job.

A week later a heavy fall raised the water level a foot. It was too wet to carry on with the dozer work so Joe turned his attention to the house. The plastered walls and ceilings had to be sanded back before painting could begin.

Kim came home with a friend and Joe gave her a fencing contract to replace the ancient original fence that was half our boundary with Bombandy; the other half had already been completed by the Bombandy workforce.

An estimate for repairs by the dam-builders was $6000, not far short of what the whole job had originally cost. Joe shook his head and patted the little dozer on its muddy tracks. For weeks he alternated between pushing and ramming dirt into the hole and undercoating house walls and ceilings.

Tracey, the guerillas' mate, came home from El Salvador in late May (one jump ahead of the Secret Police), enlivened our evenings with some black humour translated from Spanish and ran up the phone bill with rapid-fire Spanish conversations to San Salvador, couched in evasive terms in case the recipients' phones were bugged. She stayed a week before leaving for a job on a Sydney paper and left us all the more convinced that we were lucky to live in a politically stable country like Australia. A couple of days later Kim tightened up the last strain on the new fence, collected her cheque and returned to Charters Towers to paint some award-winning pictures.

On his part Joe painted too — walls, ceilings, doors and woodwork — and at long last, in August, the *House of the Thousand Cleanskins*, in all its pristine splendour, was ready for the very last move — buying the floor coverings and furniture.

For ten years I had religiously banked every cent I'd earned by writing against the day when we could spend the lot in one wild swoop. Joe and I were like a couple of kids when we set off for town. We'd never had occasion to enter a new furniture or carpet shop before; we had the money to do so for the first time.

"The lino we want, Hon," said Joe as we drove happily along, "should have a reddish background same colour as the dirt outside. That can be overlaid with brown patches and swirls and, say, little black spots here and there so when I rub up my tobacco and spill some you won't notice it and get on my back!"

"And sort of vague outlines of bare footprints in dust," I added, "so the dirt won't show up when it hasn't been swept for half a day!"

We booked into a motel and there was a carpet shop just round the corner. We walked in and our jaws dropped in amazement. On display on the back wall was a vinyl lino exactly fitting Joe's description. In addition, the shop was having a sale with greatly reduced prices. In 20 minutes we had ordered the lino, chosen the bedroom carpets, and arranged for the carpet-layer to lay them all the following week.

The nearest furniture emporium was only a street away but when we walked in a salesman came hurrying over and said,

"Sorry, that door's supposed to be closed. We're not opening today. Getting ready for the big closing-down sale next week. Then we're going into Beds Only."

"But we can't come on Monday; we're only in town for today and Saturday morning!"

He relented; he called the boss over. We explained. The boss told us to look around and if anything we liked didn't yet have the sale price marked on it, just call him over and he'd mark it down.

It was the lino story over again. The very things we wanted, of the very best quality, were there, with the sole exception of the big, glassed-in bookcase on our list. The boss and the two salespeople got almost as excited as we did, caught up in our delight. We furnished the whole house by lunchtime.

"I'd love a big round mat to match that lounge suite," I said to Joe, "but I don't suppose we'd ever find the same shade."

We went back to the carpet shop and the salesman directed us to a waist-high stack of circular carpets. The second one down was exactly right.

After lunch we ordered the making of the bookcases and on Saturday morning, with half my savings still intact, we bought a small aluminium dinghy so Joe could go fishing on the big dam. It was a fairy-tale shopping spree.

How to get it all home? We struck our first hitch trying to arrange transport. Once more Brian Moohim came to the rescue. "I'll clean out the big explosives truck and bring it out for you on Saturday week when the carpet's down," he offered.

When the time came, Jim and Carolyn came to help us unload. I'd never been more elated in my life. "One more thing," smiled Joe. "In December we make the last payment on the land and then the place is really ours!"

I proudly showed Bill and Gloria into the guest room when they came to muster. It was a far cry from when they'd first thrown out their swags on the cement floor of the old shed and shared our very spartan amenities.

We had another first that year. After three weeks of mustering on horseback Joe took up Gloria's suggestion that we try a helicopter muster to flush out the renegade cattle so adept at dodging the riders in the scrub. It was a marvellous success. Joe was instantly sold on the new method when he accompanied the pilot as spotter. The helicopter could muster in hours what it took the horsemen days to do. When at last the sale cattle were trucked and the calves branded Joe had nearly 300 weaners in the yards to be fed hay and tailed for the next few weeks so they would be properly "educated".

The other major job was the completion of the dam wall repairs. Joe had suspended work on the cave at the back because numerous swallows had nested there and he could hear the babies twittering inside the mud nests. "I'll get into it again as soon as the young ones have flown," he said, and drove the dozer across the wall to spend some days widening and deepening the by-wash.

A pair of black swans had made the dam their home, along with cormorants and nesting spoonbills and visiting pelicans. The dam was teeming with fish which, as fingerlings, Joe and I had watched swimming upstream against the overflow straight after the cyclone. When time permitted, we took the boat out and threw in a line to catch yellow-bellies and silver perch.

At the end of that year I ended the teaching career which had once fitted me like a second skin. I had by then become completely disillusioned by the farcical policies which slashed down the tall poppies among the students and reduced teaching standards to cater almost exclusively for the least competent and the least motivated. I found myself in a permanent state of frustration because the system actively prevented me from teaching anything worthwhile. The English lessons, designed to the last detail by others, were so restrictive that the poor students were even more bored than I was.

Some academic subjects (including my French and Ancient History) had long been dropped from the curriculum in favour of "Mickey Mouse" subjects, okay for hobbies but contributing nothing to employment prospects. What testing there was in English, now that examinations were taboo, was half oral — to cater for the growing number of children entering high school each year unable to read or write. Written assignments were all done at home with strict guidelines to prevent the possibility of any originality creeping in. Parents or friends could, and some often did, have too much input, so the results were sometimes totally unrelated to the student's actual ability and quite unfair to those who genuinely worked on their own.

Neither was the oral standard very high in my opinion; I reckoned that a good cattle dog had a wider vocabulary than many of the TV brain-washed students. Even some of the new breed of young English teachers, earnest and conscientious and charming as they were, would be battling to beat a really smart dog and, their own educations having been devoid of any contact with elementary grammar or the requirement to spell correctly, they were genuinely puzzled by my despairing criticisms.

I had had a gutful! I didn't want to continue an association with an education system which couldn't teach normal children to read, where streaming students was politically incorrect, where academic competition was absolutely unacceptable, and where my dissident classroom references to outmoded grammar were severely frowned upon. I was out of step with the mainstream army but, in clear conscience, I couldn't bring myself to march backwards with them any longer. On the spur of the moment I went to the office and handed in my resignation. I went home with a light heart and told Joe what I'd done.

"What took you so long, Hon!" he said.

Chapter 30

Tracey's Adventure in El Salvador.

1989 began hot and dry and Joe pumped almost non-stop from the dam to the Turkey Nest, where well over half the cattle were watering. A seepage area at the back of the dam wall towards the far end worried him. Every morning he found it trickling, but by midday it had closed off and dried back, only to re-appear by the next morning. We were told that the water level of a deep lake on a district property regularly rose and fell in harmony with the coastal tides. Was our dam moonstruck too?

"Seeping!" an old-timer dam sinker commented to me. "What big dam don't seep? Dunno any myself! Don't mean a thing! Put one down in '62, seeped then, seeping now, but she's never looked like busting through the wall." But Joe continued to monitor the seepage daily, half afraid another pipe might form.

The absence once again of a proper wet season was a more direct worry, but all fears of looming drought were reversed in the first week in April when Cyclone Aivu crossed the coast and deluged us, filling creeks and dams and sparking up the pasturage almost overnight. From then on throughout the winter we got regular storms almost as if we'd ordered them.

Our biggest problem then was the state of the road from the homestead to the mailbox and the outside world. It was no sooner negotiable than another storm put it out again. A built-up road with gravelled sections jumped high on Joe's priority list and, in the meantime, he made a lengthy detour twisting between the melon-holes at the boggiest stretch. The secret was to get into second gear and drive slowly and steadily at just the right speed for the weight of the vehicle —too slow and down you went in the mud, too fast and you slid for a hectic few moments, more than likely to end up in a water-filled melon-hole, and usually in a place where another vehicle couldn't get close enough to get a rope to your vehicle to pull you out. Joe always claimed that the way you gritted your teeth and held your jaw made the difference between getting bogged and getting through.

We left the good second-hand car I had bought with the rest of my savings out by the mailbox for most of the year and commuted to it by four-wheel-drive, tractor towing the carriage (and once by helicopter) as the state of the road decreed.

Joe and I worked out an amicable arrangement regarding a fairly common bone of contention between station and farm husbands and wives. Wives are in favour of occasional holidays together but husbands, more often than not, cannot leave the property because (a) it is too dry, (b) it is too wet, (c) it is the wrong time of the year or (d) even if it isn't any of the above, he can't find anyone he'd trust to look after the place properly in his absence.

This difference of opinion had been one of the dry creeks of our early days together when I had wept and nagged and told Joe that he might be a great lover but he was a rotten social companion. His reply had always been that I could take the kids and have a holiday any time I chose and right now would be a good time to start, so he could get a spell from my whingeing. I had long accepted that there was no way I'd ever get him to accompany me on a real holiday away; his idea of the best break he could have was a chance to paint for a couple of weeks straight when it was just too wet to leave the homestead.

Now that I was no longer teaching I was able to accept guest-speaker invitations to seminars, dinners and conventions, triggered by my newspaper column — some of them interstate, all expenses paid — and these were my holiday substitutes. I met so many interesting people, visited interesting places and my morale, which had wilted sadly in the classroom, was soon restored in my new role. Joe enthusiastically helped me to write my speeches, let me practise on him and gave me helpful criticisms, happy in the knowledge that we had at last reached a common ground that suited us both.

It was to keep a Brisbane convention appointment in July that I had to call on the helicopter pilot to ferry me to the mailbox so I could drive to town to catch the plane. It continued wet while I was away and Joe was able to indulge his urge to paint. He had painted in oils for thirty years but now he had a yen to try his hand at watercolours again. (On Mothers' Day he had disappeared for an hour or two and then re-appeared proffering me "your Mothers' Day flowers, Hon" - a delicate water-colour still-life of a bowl of flowers.) The landscape I admired on my return from Brisbane was vibrant and glowing, a scene of the western desert that was so much a part of Joe's experience. He called it *West of Alice*.

"When I've got this place ticking over properly," he said, "I'd like to go back to the big country on a painting trip. Camp out like I used to, paint a series of pictures of the old-style mustering camps, the drovers, the black ringers — the way it was in the old days."

"And the poddy-dodgers?"

"Two or three paintings for them, Hon" he laughed. "Western Queensland, the Victoria River country, top of South Australia, different strategies for different terrains." He could see those pictures in his mind — a record of a phase of Australian outback history.

Kim was painting landscapes too, among her more imaginative works, and we were able to get away together to see her second exhibition in Townsville while Bob looked after Cattle Camp for four days. We included a visit to long-time close friends, Ray and Elaine, who had recently retired to Queensland and built their dream home on the beachfront an hour's drive north of Townsville. Elaine was an artist too. It was a beaut break — a mini-holiday, like the boys' weddings.

The second half of the year was busy. Joe concentrated on his fencing program and the cattle work. We mustered by helicopter, Joe and Gloria taking turns as spotter.

Kim came with me when I drove to the Territory for a guest-speaker engagement and on my return I traded in the car for a small station wagon, brand new.

How times had changed! The old days of one beat-up old vehicle, digging postholes with crowbar and shovel, chasing snakes out of the shed and heating the bath water on the open air stove were long in the past. Joe was employing a fencing contractor, we had a nice home, a new station wagon, a sizeable cattle herd and we had paid off the freehold land and only wanted the necessary Lands Branch inspection of our improvements to finally call it our own.

We had survived depression, flood and drought and the State Government, in all our Queensland experience, had been either a National Liberal Coalition or a National Party government, both of which encouraged free enterprise and helped us through the lean years. They were a bit tardy about that inspection though; nearly a year had passed since we'd paid the last instalment on the land.

Labor won the State election in November. One of their often-repeated election promises had been that they would declare large areas as National Parks in their first year in office, which meant that large tracts of leasehold land would have to be resumed. I thought sadly of all those families, some two or three generations on the property, who would now have no security of tenure and be left wondering if they were to be among the dispossessed, their lives uprooted and their land going out of production. Queensland already had a greater area of National Parkland than other States but it wasn't as big a percentage of the State as some of the others, and to the Labor Government it was the percentage that counted.

"Thank goodness ours is freehold!" I murmured in relief to Joe.

"Well, it's not quite, yet, Hon," he corrected. "We don't get the deed until we pass the inspection!"

A week or so later the new Premier made a televised public announcement. "No more freehold land will be granted in Queensland!" We were eating tea and watching the news when his announcement was made. We both went white with shock and stared at each other. Had all our effort been for nothing? Without freehold we had no security and our asset would plummet in value! We pushed away our plates and stood up. Joe put his arms around me and held me tight. The newsreader rattled on but we didn't hear a word he said.

We discussed the situation over breakfast the next morning.

"It has to be a qualified statement," I reasoned. "He can't possibly mean that city people in the future can't buy a freehold block on which to build a house."

"And country property that is already freehold should be safe enough."

"Surely we'll be all right, seeing we're paid up, but I wonder what the score will be for the ballot blocks. With 25 or 30 years to pay off the land, they've still got a way to go."

"Well," said Joe, "let's not cross the creek before we come to it. I'm off to cut some strainer posts. Try not to worry about it, Hon."

Easier said than done. At midday I phoned the District Lands branch manager. "The phone hasn't stopped ringing all morning," he told me. "I'm sorry, I don't know any more than you do. It's a bit chaotic down in Head Office. I can't get a clear definition on anything. There's staff changes and transfers going on. About your inspection. I'll get back to you as soon as I can."

The District staff must have worked overtime because their phone calls after that were all made about 9pm. The inspection was arranged and finally carried out. "The paperwork's all gone to Brisbane," we were told, "but don't expect results too soon. Still no clear guidelines down there. All I can say is that your inspection passed muster and the grant has been recommended." With that we had to be content.

Our Christmas was the usual family gathering with the exception of Tracey, who had returned to El Salvador, freelancing again.

Jim and Carolyn had saved assiduously and planned to leave on an extended working holiday overseas. They brought their household furniture and personal gear and stored it in the old shed.

The weather continued hot and dry and once again Joe's main occupation was pumping water to keep up with the demands of thirsty stock. Because of the winter rains of the previous year there was still a good body of dry feed but, with the wet season rains overdue, that in itself presented a problem. Bushfires! Each day we noted the grey smoke columns rising on one or the other horizon and Joe anxiously widened our fence-line firebreaks.

Three days after Christmas our turn came. The fire swept through from Willunga, raced across all our frontage country, roared into Lilianvale and then jumped across the main road and into Seloh Nolem. All the men could do was cut firebreaks and burn back to the fire. Neighbours worked together in a desperate bid to save at least some grass and at night they doused burning tree stumps in case an errant gust of wind snatched up the coals and started a new blaze.

A week later another fire jumped across the firebreak on our northern Amaroo boundary where the fencing contractor was working. With his two-way radio he quickly alerted the Amaroo owner and his neighbours who were working on another face of the fire and they arrived within minutes. In the meantime Joe hastily graded another firebreak with the little dozer. With so many workers, each with a water tank on the back of his vehicle, that fire was soon contained.

It is the nature of the bushman to face a setback and turn it to his advantage where he possibly can. With all the frontage country now a bed of white ash, Joe's mind leapt ahead to picture it as a sea of waving pasture grasses. He ordered the bags of seed and Brian Moohin brought them on a semi-trailer, unhitched and left the tarpaulin-covered trailer on the Lilianvale airstrip, which was much closer to the burned area than our own strip some miles away.

The pilot of the small aerial-seeding plane was a lot older than the usual run of young pilots but he had that devil-may-care attitude that instantly reminded us of the war-time pilots with their handle-bar moustaches. Viv brought two "kickers" with him. His seeding method was simple. Joe and one of the kickers crammed as many bags as would fit into the body of the plane and the other kicker crawled in on top of them. Viv took off and flew low while the kicker slit open bags and kicked their contents out through a hatch. Each sweep was about five miles long and the kicker was only too pleased to stagger out at the end of the run and let the other kicker take his place for each alternate run. It was stinking hot and the tiny seeds stuck to clothes and hair and sweat-stained faces, yet the men grinned and chiacked each other and stuck to the job for the full day and half the next day. If anyone ever earned their pay, those kickers did. Joe estimated that we had seeded between eight and ten thousand acres with a mixture of half buffel grass and a quarter each of Green Panic and Rhodes grasses.

I don't know how much of the seed actually stayed where it fell in early January. It was late March before cyclone Ivor finally brought us good rains to germinate the seed, and every day in between, at any given time, you could watch six or seven white-ash whirlywinds marching their columns across the burn, zipping across the main road, spinning into the adjacent scrub. All I know is that, outside our fence on either side of the main road, there now grow stands of lush buffel grass that would gladden the heart of the most finicky cow. Inside the fence isn't too bad either, so some of the seed must have stayed. It was a lot better result than our first seeding years before when we borrowed $10,000 to pay for the seed and got two inches of rain to germinate it a fortnight later, then watched for hot, dry week after week as it all withered up and died away. Joe seeded that lot himself, too, because we couldn't afford the plane, driving up and down on the dozer for days with his seed-scattering invention dangling from the dozer's tree-pusher attachment. Pasture-seeding is like everything else in the bush; no matter how you plan, that wild card, fickle nature, controller of rainclouds and whirlywinds, really calls the tune.

The work pace hotted up. Joe cut and carted strainer posts one jump ahead of the fencing contractor. He ministered periodically to the troublesome generator, new but a lemon right from the start, installed a new engine and pump at the dam, built a carport for the station wagon and carried out the general station maintenance.

There were two long days, however, the 29th and 30th of April, when neither of us gave the slightest thought to any property demands.

About 1 a.m. on the 29th the insistent ringing of the phone awakened us. When I sleepily picked up the receiver a female voice with an American accent said quickly,

"Tracey's been arrested by the Treasury police. You must phone the Police Chief before she "disappears", so that he knows that you know who took her. His name is Pedro Gomez and the number is ---."

I wrote down the number by torchlight and asked,

"Does he speak English?"

"No!"

"Who's speaking?" I stammered.

"Sorry, I can't say!" The phone went dead.

Joe rushed over to the engine shed to start up the lighting plant. He lit the fire to make a cup of tea while I struggled frantically to construct the right sentences in my very limited Spanish. I got through to Senor Gomez in San Salvador straight away but neither of us understood more than a fraction of what the other was saying for the next twenty minutes. However, at least he knew that Tracey's parents knew her whereabouts and would pass details to the Australian Government. When I demanded to know why she was there, all I could understand of his answer was something about "not holding" but I wasn't sure whether he was denying holding her in custody or whether she had been arrested because she wasn't holding the right papers. It turned out to be the latter.

The American girl who had alerted us, had first contacted the Reuters correspondent, who had arrived at the house in time to photograph Tracey being taken away at gunpoint. He followed the police jeep to a suburban police station and then contacted the other foreign correspondents. They all gathered at the police station and refused to leave until she was released.

Meantime, Joe phoned the Foreign Affairs Department in Canberra. Despite the time — the early hours of Saturday morning — the personnel there swung into action immediately. They were wonderful. Australia doesn't have a Consul in El Salvador but, working through our Consul in Mexico City and the British Consul in San Salvador, they secured her release in three hours.

Tracey phoned us with a hasty explanation. She said she had about six hours before the possibility of re-arrest. In that country foreign journalists must have a press accreditation from their papers, renewable every six months. Tracey's had allegedly not been renewed but I knew that both *The Australian* and *The Canberra Times* had faxed her accreditation to the Treasury officials in mid-March because she had phoned me then to contact both papers to send it again when the officials kept denying that any notification had arrived.

Hastily, she told me that the *New York Times* correspondent had a fax machine in his hotel bedroom. She gave me his fax number and asked me to ask *The Australian* and *The Canberra Times* foreign editors to send the accreditations once again, but this time to the hotel-room fax machine. They quickly did so. *The Australian* editor said that it was the fifteenth time she'd sent it.

A few hours later Tracey fronted up at the police station with her faxed credentials from both papers. By this time the Reuters photo of a soldier holding a machine gun in her back was appearing in newspapers all round the world. The senior official who stamped her permit was very apologetic. He showed her a whole stack of faxes which he claimed "have only just arrived today!". The first one was dated March 11th. With Latin charm he offered to give her a 12 months' permit in compensation instead of the usual 6 months. He asked her to write a report of the incident and returned her confiscated notebooks. The five rolls of film they had taken were being developed for her at no cost, he said. Tracey was wild about that because her particular brand of film couldn't be developed in San Salvador and the five rolls were consequently ruined when they tried. "But," she told me much later, "they didn't find what they were really looking for!"

At the time of her arrest soldiers, armed with machine guns, hand-grenades and knives, had held her for five hours while they searched the house where she roomed from top to bottom, looking for "documents of illegal women's organisations".

"I was pretty lucky really, Mum," she told me. "I happened to glance out of my upstairs window and I saw the jeep loaded with soldiers turn the street corner. (A few weeks earlier she and a French journalist had been grabbed in the street, bundled into a car, blindfolded and driven around for three hours, before being thrown out on the city's outskirts. She had no doubts that this vehicle was coming to her house.) "Right under my window," she continued, "an open-back rubbish truck was parked while the rubbish-men collected the bins from the houses around. I just dropped the computer disc I didn't particularly want them to find straight from my window into the middle of the rubbish and watched the truck lumber away, just as the jeep drew up!"

When all the furore died down Joe and I suggested to Tracey that it might be a good idea to leave El Salvador. The British Consul had told her to phone in daily and let him know where she was going each day. Obviously he too believed that she was not safe. "Yes," she agreed, "I've already decided to leave for a while. I'm going to Guatemala next week to cover the elections there!"

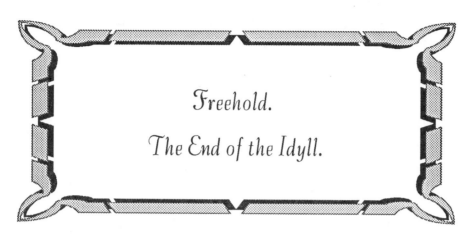

Chapter 31

Freehold.

The End of the Idyll.

It was a cold winter that year. The best part of the day was the first half-hour when we sat companionably in front of the kitchen fire, drinking coffee and discussing short and long term plans, while we waited for it to get light. By mutual agreement we didn't mention the word freehold.   There had been no further public announcements on that subject.

The one thing that did puzzle us, though, was the Premier's interview on TV when he denied that any leasehold station properties had been resumed for National Parks. We personally knew a family on the Barcoo, fourth generation on their property, who had been given 12 months to leave the property and had been promised compensation for its improved value, so they could buy another property and move their stock.   They had found a suitable property for sale but the compensation payment hadn't been forthcoming in time so the vendor had sold elsewhere. We'd heard of three other properties up the Cape in similar straits.

Each winter Joe suffered a few days or weeks of painful joint swelling but this time it was so persistent that I finally persuaded him to see a doctor. He hadn't had any contact with a doctor since the prawn incident and his "grin and bear it and it will eventually go away" attitude so worried and frustrated me that I performed and carried on until he gave in and let me drive him to town. Blood tests revealed that a recurrent Dengue Fever virus was the problem and medication immediately relieved the joint pains.

Next morning, sitting in front of the fire after the first pain-free night for ages, he brightly said,

"Y'know, Hon, I should have gone to see about this years ago!" When I thought of all the times I'd nagged him to no avail and been snapped at for my pains, I had to control a wild impulse to push him off his chair and kick him in the ribs. I couldn't safely comment so I patted his arm instead.

With renewed vigour he set about the extensions to the homestead yards that he wanted to complete before the serious cattle work began.

One day in the middle of June I woke up with the thought of the freehold title in the forefront of my mind. It kept intruding into all my other thoughts as the morning wore on. Finally, I telephoned the District Lands Department manager. "Haven't any confirmation yet, Marie," he told me, "but, look, I'll phone Brisbane and I'll ring you back later."

Joe had just walked into the kitchen for afternoon smoko when the phone rang. "You're going to like this," said the familiar voice. "The governor signed your Deed of Grant yesterday afternoon. It's yours! You'll get your official confirmation in the mail." I beamed at Joe and relayed the great news. He just stood stock-still with a funny little smile on his face and said "Well, well!".

We had a prolonged smoko that day. From early dreams to final culmination and such a near thing too! We rejoiced in the good fortune that had made it possible. Thirty-eight years from two quid, a swag and a dream to a cattle station in our own names! Joe had been the driving force; his were the plans and most of the physical hard work and all the responsibility. I had only tagged along as a backstop. The kids had been marvellous with all their help over the years.

Joe went off to start the pumps and I phoned Kim and Bob but as Tracey and Jim were both overseas, we had to wait until they phoned us to pass on the good news.

It wasn't so much the fact of owning a station that made us so happy; it was having security for our way of life that counted. With freehold title we couldn't be forced off against our wishes as easily as might happen with leasehold titles. Joe's affinity with the bush and its creatures was so strong that he didn't fit comfortably in any other environment. The bush was his element, the wilder the better. Loving the life as I did, it had still taken me a long time to understand this basic and primitive need in my mate.

I knew there would be certain scrub areas on Cattle Camp that would never be disturbed in his time. He didn't aspire to use all the land to its full stock potential, to treat it as a business venture only. "Nobody really owns the land, it's just on paper!" he told me. "As I see it, we're just caretakers while we're here. We share it with nature. If we don't, we'll be the losers in the end!" We shuddered at the thought of the already over-populated world and acknowledged that we were among the most privileged of humans, still able to pursue a way of life in Australia that had long disappeared in most other countries.

I allowed my romantic streak to surface. I thought - Hiawatha talked to the animals and birds of the forest; St. Francis of Assisi did too; and so does Joe. I'd seen his affinity often enough — the wild budgerigars in the desert alighting on his shoulder and clinging to his shirt as he walked; the little organ-grinder lizards exploring his proffered hand; wallabies and kangaroos standing, heads cocked, as he approached and then dropping, re-assured, to feed again; the way hurt creatures quietened and responded to him when he tended their injuries. Our way of life was something you couldn't put a price on; it was the sweet icing on the prosaic damper of hard bush toil.

Joe finished the new wing of the yards and Bob, now working for himself, arrived one Sunday in late August in his truck with his new four-wheeler bike on the back. He and Joe spent a few days doing last-minute pre-muster jobs and yarned together each evening. Our usual helicopter pilot had returned to New Zealand and the new pilot was booked for 8 o'clock on Thursday morning, timed to coincide with Bill and Gloria's arrival.

Joe and Bob went down early to the yards. Gloria phoned me that they had been delayed, so when the chopper landed at the house I re-directed the pilot to the yards and asked him to pass on the message. Then I started in on the extra cooking; I wanted to get the meals organised ahead as far as possible so I could be on hand to help with the tallying when the cattle were in the yards.

Gloria called in about eleven.

"Sandy's two-way radio in the chopper must have packed it in," she grumbled.

"I can't contact him at all and we don't know where they were going to muster first. Bob's waiting down at the yards with Bill. Unless we hear the chopper soon, we'd better go scouting around!"

Dinner time came and went. Then the phone rang and a voice said, "Nebo Ambulance station here. The ambulance is on its way!"

"The helicopter?" I stammered, "Who's hurt?"

"The pilot is badly injured ----------". I dropped the phone as Bob burst through the door and grabbed hold of me.

"Dad's dead!" he said quietly, irrevocably.

He phoned Kim. The Nebo ambulance broke down. The Dysart ambulance took a wrong turning and got lost. Bill stayed with semi-conscious Sandy, still trapped in the wreckage. Gloria drove to the nearest phone at Willunga when the hours passed and no ambulance arrived. The Middlemount policeman came and asked for the loan of a chainsaw. An ambulance finally arrived just before dark and the Nebo policeman stayed to camp overnight at the crash site until the arrival of the Bureau of Air Safety Investigation team.

Kim came an hour after midnight with the friend Bob had insisted she get to co-drive with her. It was after two o'clock before we all went exhausted to bed. Gloria, efficient Gloria, had organised everyone. Until then I hadn't been left on my own for a moment to think.

I had always puzzled that anyone could ever contemplate suicide. Life was so precious and nothing could ever be so bad that the future wasn't worth a chance. Now I knew differently. There was no future without Joe. I couldn't, didn't want to, go on without my mate. I lay awake in despair.

Then, suddenly, he was right there, standing beside the bed. I jumped up. His face wore that funny, rueful little smile that had always tugged at my heart. He put one hand on my shoulder and looked straight at me.

"Buck up, Hon!" he said firmly. He smiled again and then he left me.

I did sleep then, only to be awakened suddenly in the pitch dark by a volume of sound, a crescendo of wild, eerie keening that jolted everyone in the house wide awake. Just beyond the garden fence hundreds of curlews were gathered, crying and calling together in a tumultuous torrent of mourning, a wild requiem that washed across the homestead, echoed into the gully and rose into the wide night sky. They called for perhaps a quarter of an hour, then the last pure high note died away and the bush was silent again.